SIR P

THE SKY
AT NIGHT

2001-2005

My grateful thanks go to Daniel Clarke, who took my somewhat chaotic manuscript of this book and produced an impeccable typescript.

Selsey, 6 June 2005

First published in Great Britain in 2005
by Philip's, a division of Octopus Publishing Group Ltd,
2–4 Heron Quays, London E14 4JP

ISBN-13 978–0–540–08808–9
ISBN-10 0–540–08808–0

Printed in China

Details of other Philip's titles and services can be found on
our website at: **www.philips-maps.co.uk**

FRONT MATTER IMAGES
Right: Stephan's Quintet, as
imaged by the Gemini
Observatory.
Pages 4–5: V838
Monocerotis, as page 8.
Page 6: Very Large
Telescope, photographed by
Chris Lintott.
Page 8: Red supergiant star
V838 Monocerotis, imaged
by the Hubble Space
Telescope.
Page 11: My BAFTA (British
Academy of Film & Television
Arts) special award.

SIR PATRICK
MOORE
THE SKY
AT NIGHT
2001-2005

CONTENTS

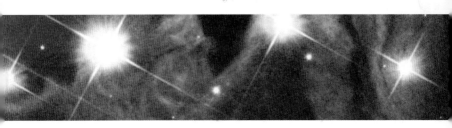

ACKNOWLEDGEMENTS

My grateful thanks to those who have joined me on *The Sky at Night* during this period. In chronological order:

Chris Lintott	Phil Dianoni	Brian May
John Mason	Peter Wilkinson	Colin Pillinger
David Hughes	Danielle Kettle	Monica Grady
Mark Kidger	John Dobson	Garry Hunt
David Wynn-Williams	Steve Wainwright	Donald Francke
Chris Kitchin	John Fletcher	Neil Crossland
Douglas Arnold	Martin Rees	Allan Chapman
Fred Watson	Iain Nicolson	Catherine Galloway
Stephen Hawking	David Hardy	Peter Cattermole
Duncan Steel	Tony Wilmot	Andrew Coates
Bernard Lovell	Carlos Frenk	Derek Ward-Thompson
Andrew Lyne	Gordon Rogers	Mike Bode
Ian Morison	Nial Tanvir	Martin Mobberley

PHOTOGRAPHIC ACKNOWLEDGEMENTS

'PM' indicates the Sir Patrick Moore Collection
Front/back cover NASA, ESA and A. Nota (STScI/ESA); 3 Gemini Observatory/Travis Rector, University of Alaska; 4–5 NASA and the Hubble Heritage Team (AURA/STScI); 6 Chris Lintott; 8 NASA and the Hubble Heritage Team (AURA/STScI); 11 PM; 12 NASA/JPL; 13, 14 PM; 16 Eyad Mustafa; 19 Scrovegni Chapel, Padua, Italy/Bridgeman Art Library; 20 PM; 22 Philip James (University of Toledo)/Steve Lee (University of Colorado)/NASA; 23 ESA/DLR/FU Berlin (G. Neukum); 24 Steve Lee (University of Colorado)/Philip James (University of Toledo)/ Mike Wolff (University of Toledo)/NASA; 25 G. Neukum (FU Berlin) et al./Mars Express/DLR/ESA; 27 Bill Schoening/NOAO/AURA/NSF; 28 Travis Rector (NRAO/AUI/NSF and NOAO/AURA/NSF)/M. Hanna (NOAO/AURA/NSF); 29 Travis Rector and Monica Ramirez/NOAO/AURA/NSF; 32t PM; 32b John Fletcher; 33 Gary White/Verlenne Monroe/Adam Block/NOAO/AURA/NSF; 34 AURA/NOAO/NSF; 37 NASA; 38 Edwin Aldrin/NASA; 39 Neil Armstrong/NASA; 41 NASA/JPL; 42 Malin Space Science Systems/MGS/NASA; 43 NOAO/AURA/NSF; 45, 46, 47 PM; 49 NASA/ESA/Richard Ellis (Caltech)/Jean-Paul Kneib (Observatoire Midi-Pyrenees, France); 50 DMR/COBE/NASA/Two-Year Sky Map; 51 Rogier Windhorst and Sam Pascarelle (Arizona State University)/NASA; 52 NASA, the Hubble Heritage Team and A. Reiss (STScI); 55 John Fletcher; 56 PM; 57, 59 John Fletcher; 60 NASA/GSFC/LARC/JPL/MISR Team; 61 NASA/JPL; 62t John Fletcher; 62b NASA/GSFC/METI/ERSDAC/JAROS and US/Japan ASTER Science Team; 63 John Fletcher; 64 Ian Morison, Jodrell Bank Observatory; 65 Courtesy European Southern Observatory; 66 L. King (University of Manchester)/STScI/NASA; 68 Courtesy of the NAIC–Arecibo Observatory, a facility of the NSF; 71 Courtesy European Southern Observatory; 72 High-Z Supernova Search Team/HST/NASA; 75, 76, 79 Courtesy of SOHO/MDI Consortium. SOHO is a project of international cooperation between ESA and NASA; 80 Royal Swedish Academy of Sciences; 88 NOAO/AURA/NSF; 89 the Boomerang Collaboration; 91 Courtesy of SOHO/MDI Consortium. SOHO is a project of international cooperation between ESA and NASA; 92 John Fletcher; 94, 95, 96 PM; 98 A Fruchter (STScI)/NASA; 99 Courtesy European Southern Observatory; 100 Cenko, Soifer, Bian, Desai, Kulkarni (Caltech)/Berger (Carnegie)/Dey, Jannuzi (NOAO); 102–3 S. Beckwith (STScI), Hubble Heritage Team (STScI/AURA)/ESA/NASA; 104 NASA/H.-J. Yan, R. Windhorst and S. Cohen (Arizona State University); 105t Rogier Windhorst and Sam Pascarelle (Arizona State University)/NASA; 105b WMAP Science Team/NASA; 106, 107 NASA/H Ford (JHU), G Illingworth (UCSC/LO), M Clampin (STScI), G Hartig (STScI)/The ACS Team/ESA; 108 ESA/SPACE-X/AMIE TEAM; 109t ESA; 109b ESA (illustration by Medialab); 110 ESA; 112, 113 NASA/JPL/University of Arizona; 114 NASA/JPL/Cornell; 115 NASA/JPL/University of Arizona; 116, 117 John Fletcher; 118 NASA/JPL/Caltech; 119 NASA/AURA/NSF; 122 PM; 123 Chris Lintott; 124 PM; 126, 127 Chris Lintott; 128, 128–9 NASA/JPL/CORNELL; 131 Ian Spooner (University of Sheffield); 132t, 132b Alan Schultz; 135 Courtesy European Southern University; 136 Brad Whitmore (STScI)/NASA; 138 Courtesy European Southern Observatory; 140 Jamie Cooper; 141 Jamie Cooper/VT-2004 programme; 142 Martin Mobberley; 143l, 143r John Fletcher; 145, 146, 147 NASA/JPL/Space Science Institute; 148 Hubble Heritage Team (AURA/STScI)/NASA; 149, 150 NASA/JPL/Space Science Institute; 152, 153 Chris Lintott; 154 Faulkes Telescope Project/Nik Szymanek; 155 Chris Lintott; 157 Raghvendra Sahai and John Trauger (JPL)/WFPC2 Science Team/NASA; 158–9 NASA/ESA/C.R. O'Dell (Vanderbilt University)/M. Meixner and P. McCullough (STScI); 160 Courtesy European Southern University; 161 Bruck Balick (University of Washington), Vincent Icke (Leiden University, the Netherlands), Garrett Mellema (Stockholm University)/NASA; 163, 164 Chris Lintott; 165 PM; 166–7 Courtesy European Southern Observatory; 169, 170–1, 171 ESA/NASA/JPL/Descent Imager/Spectral Radiometer Team (LPL); 173, 174 John Fletcher; 176, 177, 178 Pete Lawrence.

Artwork acknowledgements
69, 73 Raymond Turvey/Philip's; 82 Paul Doherty; 83 Frank Norrington; 85tl Christopher Harrison; 85tr Jeffrey John; 85b Dan Bright; 86 Gertrude L. Moore/PM.

FOREWORD

The Sky at Night 2001–2005

Writing the foreword to any book is a daunting task. This is a challenge I readily accepted, however, because as a youngster in the 1950s I watched Patrick's early appearances on children's TV and was captivated by his humour and excitement as he imparted to us a new subject – astronomy.

The British public are very fortunate. As scientists gain greater knowledge of the night sky, and the planets in our Solar System are explored by sophisticated NASA and ESA space missions, we are informed immediately, in the warmth and comfort of our own homes, through the regular monthly editions of *The Sky at Night*. Since 1957, Patrick Moore, with his boundless and youthful enthusiasm, enormous astronomical knowledge and array of leading astronomers and space scientists from all over the world, has kept us informed of every new discovery and development in astronomy, in a manner we can all easily understand. This is no mean feat for a topic that so many would consider beyond them. But Patrick is special; indeed, I am prepared to say that Patrick Moore is unique. Already a legend in his own time, he has been described by a past Head of my College as '… more valuable to astronomy than most of the Astronomers Royal in history…'. Patrick is respected by the world's astronomical community and the international space agencies, and this book chronicles the major astronomical developments of the past three and a half years, which we have been able to share with Patrick and his distinguished guests.

During the course of this set of programmes, we are taken from participating in the activities of a local astronomical society to sharing in the breathtaking results from the Mars missions and the Huygens landing on Titan. We also share in discussions of the current astronomical theories which are pushing back the frontiers of our knowledge. The real joy of the book, however, is that every topic, no matter how complicated the title may first appear, is made so readable and easy to understand.

This book is more than just a record of another three and a half years of *The Sky at Night*, the longest-running TV programme. It is also a record of the enormous debt we owe Patrick for maintaining his enthusiasm for keeping the public informed. Thank goodness this great talent has been recognised. In 2001, Patrick was made a Fellow of the Royal Society and received a Knighthood. He also received a BAFTA award for his untiring activities and special ability for communicating astronomy to everyone, from the very young to the more mature, whatever their background, in a manner they can understand.

This book is full of fascinating information, written imaginatively to excite and stimulate the reader; it is a wonderful record of an invaluable set of programmes.

Professor Garry E Hunt
March 2005

INTRODUCTION

This collection of articles based on *The Sky at Night* programmes – the 12th book in the series, though only the second to be published by Philip's – covers the period from late 2001 to early 2005. It follows the same pattern as in the earlier books, so that we have a variety of topics; some of the articles are more technical than others, but I hope that there is 'something for everybody'.

I am most grateful to all those distinguished astronomers who have joined me in the programmes; I have listed them on page 6. Meanwhile, there is one point I must make. In most of the programmes given here, I have been joined by guests, and inevitably the texts in the present book are based upon these discussions, though of course any errors and omissions are my responsibility alone.

Much has happened since that first *Sky at Night* programme, almost 50 years ago. And I am sure that there will be plenty more to say when I start to compile *The Sky at Night* book, issue 13!

Patrick Moore

Sir Patrick Moore
Selsey, July 2005

1 • THE LEONID STORM

Comets are the most erratic members of the Solar System. Flimsy and fragile, they were once regarded as 'flying gravel-banks', but are now more properly called 'dirty ice balls'. The only substantial part of a comet is its nucleus, which is never large. Comet Hale–Bopp, the beautiful comet which ushered out the 20th century, had a nucleus no more than 40 km (25 miles) in diameter – which makes it a giant by cometary standards.

Until 2001 the only cometary nucleus which had been seen close-up was that of Halley's Comet, encountered in 1986 by the spacecraft Giotto. When a comet nears the Sun, its nucleus is hidden behind a cloud of dust and gas, so that space missions give us the only opportunities to see it.

A chance came on 22 September 2001, when the US probe Deep Space 1 flew past Borrelly's Comet and sent back excellent pictures. Borrelly is a modest comet compared with Halley; it has an orbital period of just under 7 years and has been seen regularly since its discovery in 1904, though it never becomes bright enough to be seen with the naked eye. Deep Space 1 swooped past the nucleus at a range of 2200 km (1350 miles), and the camera worked excellently; at this stage the comet was 220,000,0000 km (137,000,000 miles) from the Earth and 204,000,000 km (127,000,000 miles) from the Sun. The nucleus was shown to be elongated, 8 km long by 4 km wide (5 by $2\frac{1}{2}$ miles), and decidedly complex, with some smooth areas and rugged regions, with jets of gas and dust shooting out. But the real surprise was that the nucleus was so dark – in fact, the darkest body in the Solar System. The surface seems to be coated with a charcoal-like substance, less reflective even than the surface of the Moon.

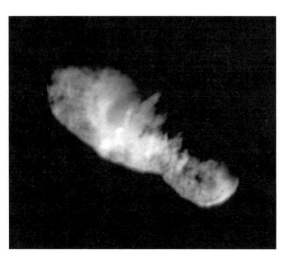

◀ **Comet Borrelly's nucleus,** taken from Deep Space 1. This image was taken just 160 seconds before the spacecraft's closest approach to the comet.

As a comet moves through space, it leaves a dusty trail behind it. When the Earth passes through such a trail, some of the particles crash into the upper air. Moving at up to 72 km (45 miles) per second, they collide with air particles and are heated by friction, so that each particle burns away to produce the luminous streak that we call a meteor or shooting star. What we see is not the particle

▲ *The Leonid Shower of 1833* was depicted in the 1889 edition of Bible Readings for the Home Circle. At the storm's peak, tens of thousands of meteors were seen per hour.

itself, which is of sand-grain size, but the effects in the atmosphere due to its suicidal plunge. Meteors burn out at around 60 km (40 miles) above sea-level; they are quite unlike crater-producing meteorites, which come from the asteroid belt and are in no way associated with comets.

There are many annual meteor showers, many of them linked with known comets (though we do not pass through the Borrelly trail). Sporadic meteors, not connected with showers, may come at any moment. Each August we have the Perseid shower, associated with comet Swift–Tuttle, and in December we have the Geminids; with rich showers the ZHR may exceed 80. (The ZHR, or Zenithal Hourly Rate, gives the number of naked-eye shower meteors which would be expected to be seen by an observer in ideal conditions, with the radiant at the zenith or overhead point. In practice, these conditions are never fulfilled, so that the actual number of meteors seen is always lower than the theoretical ZHR.)

The November meteor shower has its radiant in the constellation of Leo, the Lion, and is therefore termed the Leonid shower. Unlike the Perseids and the Geminids, the Leonids are not consistent. The parent comet – Tempel-Tuttle – has a period of just over 3 years, and passes close to the Earth's orbit only three times per century; the last occasion was in 1998.

▼ **The constellation of Leo,** *as it appears in* Uranometria, *the star* *atlas by Johann Bayer,* *published in Augsburg,* *Germany, in 1603.*

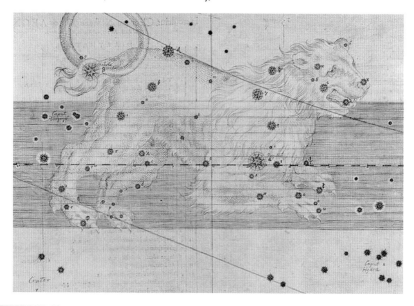

Each time the comet goes round the Sun, it produces a new trail of dusty debris. Initially, the dust in these narrow trails is highly concentrated, but as time goes by the dust spreads out, and eventually disperses over the whole of the comet's orbit. The Leonid meteor stream is young by astronomical standards, and there has not been enough time for it to have spread along the complete orbit. This uneven distribution of dust means that although we see a few Leonids every year, the most spectacular displays are confined to the few years before and after the comet passes closest to the orbit of the Earth. If on these occasions the Earth passes through, or close to, one of the narrow filaments in which the dust is concentrated, we see a veritable 'meteor storm'.

The first such 'storm' to be recorded was that of the year 902, seen from southern Europe and northern Africa. Others followed at intervals of around 33 years, but the records are fragmentary. The first detailed records date from November 1799, when a graphic account was given by two observers in South America, the German explorer Alexander von Humboldt and his French companion Aimé Bonpland. Even greater was the display of the 12–13 November 1833; this time it was best seen from the eastern United States, with meteors falling at a peak rate of tens of thousands per hour. In November 1866 there was another Leonid storm, well seen from Britain and Europe. The meteor rates were not so high as those of 1833, but the maximum observed rate was around 120 meteors per minute, or 7200 Leonids per hour. The maximum activity lasted for no more than an hour or two, and the actual peak was limited to only a few minutes.

There were no meteor storms in 1899 or 1933, because the narrow dust filaments did not pass sufficiently close to the Earth, but on 17 November 1956 the Leonids were back, producing a storm comparable with that of 1833. Once again the peak was very narrow, lasting for just under an hour, and it occurred during daylight in Britain, so that little was seen; observers in the United States were treated to a truly magnificent display, with maximum rates of well over 100,000 meteors per hour.

We produced a special *The Sky at Night* programme, and we issued over 10,000 report cards, hoping that the predications were wrong and that the storm would after all be visible from Britain. Viewers co-operated nobly, but meteors were sparse. I still have one card sent to me by a viewer in Hampshire: 'Watched sky from midnight till dawn. Meteors – from the sky, none. From the wife, plenty!'

Tempel–Tuttle returned once more in 1998, and meteor observers all over the world were on the alert. Enhanced Leonid activity was noted as early as 1994, and picked up as the comet closed in. There was no storm in 1998, but observers in Europe did see a very fine shower of bright fireballs about sixteen hours before the predicated peak of the main shower. These fireballs were due to relatively large

◄ *The Leonid Storm of 1999,* seen in superb conditions in the desert of eastern Jordan. More than 70 meteors are on the original negative of this 10-minute exposure.

dust particles, up to a few centimetres in diameter, which had been ejected from the parent comet during the return in 1933.

The distribution of dust particles around the comet means that the best Leonid storms occur after the comet has passed closest to the Earth's orbit, not beforehand. There were many predications about what would happen in 1999; in particular there was some interesting work by David Asher (Armagh Observatory) and Robert McNaught (Australian National Observatory), who modelled the narrow dust trails laid down by the comet over many returns, and worked out their positions in space relative to the Earth. In 1999 they found that the Earth would be near the 1866 and 1933 trails shortly after midnight on 18 November. They were surprisingly accurate. The best views were obtained east of Britain: John Mason, in the Sinai Desert, recorded a peak rate of 2500 meteors per hour. Again I was unlucky: from my Selsey home I sat discontentedly in my garden, together with the BBC team, holding an umbrella to protect myself against the raindrops falling gently from overhead. At Oban, in Scotland, Iain Nicolson was blessed by better fortune; he saw a brief but spectacular shower, with a peak ZHR of over 2000. Many of the meteors were brilliant, with long, sometimes persistent trains.

Incidentally, it is worth noting that the comet and its meteors move round the Sun in a retrograde direction, so that we meet the Leonids head-on. Moreover, the orbit of the comet is inclined to that of the Earth by $17\frac{1}{2}$ degrees.

In November 2000 the Leonids were fairly rich, but there was no real storm, and observations were badly hampered by moonlight. Asher and McNaught were again very accurate. The prospects for 2001 seemed to be better, and the Moon would be out of the way.

As usual, we made arrangements for television coverage, but it was clear that Britain would not be a good site; the main display would occur when the Leonid radiant was still below our horizon. John

Mason went to an island in the Pacific, and was well rewarded. At Selsey, I was once again totally clouded out.

As usual, the maximum period of activity was limited, but in Australia some observers reported three hours of beautiful Leonids, many with orange or yellow heads and blue-green trails. There were frequent fireballs, some of which outshone Venus and Jupiter, and at one stage the meteors were so numerous that counting them was impossible.

When will there be another Leonid meteor storm? Perhaps not for many decades. Comet Tempel–Tuttle is due back once more in 2031, so that we may be lucky in the period from 2032 to 2034, and no doubt *The Sky at Night* will be ready, but I am afraid that there will be another presenter. I do not expect to live until the advanced age of 109, so I will not be able to see the Leonids at their best. I will have to be content with the Perseids instead.

— 2 · *THE STAR OF BETHLEHEM* —

In December 2001 the planet Venus was at its best. It is far brighter than any other object in the sky apart from the Sun and the Moon, and it is capable of casting perceptible shadows. As Christmas approached, the inevitable question was asked: could Venus be the Star of Bethlehem?

Unfortunately, our information is deplorably scanty. The Star is mentioned only once in the Bible, in the Gospel according to St Matthew 2.1–12, 16. None of the other Gospels says anything about it, and there are no references anywhere else. Even St Matthew is infuriatingly vague. He merely says the Star was seen in the east, that it 'went before them' until it came over the place where the young Jesus lay, at which point it stopped. What happened to it afterwards we do not know. Moreover, we are very unsure of our timescale. There is no doubt that Jesus was born well before AD 1, and we are limited to the period between BC 7 and BC 9; neither did 25 December have any special significance until several centuries later – by which time the real date had been forgotten, so that our Christmas is wrong too.

For this special *The Sky at Night* programme I was joined by two old friends, Professor David Hughes of Sheffield University and Dr Mark Kidger of the observatory at Tenerife. All three of us had written books about the Star, and each of us had come to different conclusions. But before starting the discussion, I made two points. There are many people who believe that the whole story was merely invented by St Matthew as a nice touch. Others maintain that the Star was supernatural. We did not intend to go down either of those roads. All we set out to do was to see if there were any astronomical phenomena which would provide an explanation – assuming that the star had really been seen.

Many ideas have been proposed, but most can be dismissed without further ado. The Wise Men – the Magi – were certainly astrologers with an excellent working knowledge of the sky, so they would not have been deceived by Venus or any other planet or 'normal' star. Had a brilliant comet appeared at the time of the nativity, it would have been mentioned by contemporary astronomers (incidentally, the only bright comet which is seen regularly – Halley's – returned several years before BC 7, the earliest date when Jesus could possibly have been born). The same objection disposes of a supernova, a colossal stellar outburst in which a formerly dim star literally blows itself to pieces. The last supernova to have been seen in our Galaxy flared up as long ago as 1604, and there is no record of anything of the kind for many years to either side of AD 1. Aurorae, or Northern Lights, are very rare indeed at the latitude of Bethlehem, and in any case it is inconceivable that an aurora could be confused with a star.

You have to look for something that was unusual, was conspicuous, was seen only by the Wise Men, and which moved in a way quite unlike that of any normal star or planet. And we have to admit straight away that there is no object which can meet all these requirements – apart from a flying saucer, which I for one do not consider very likely!

David Hughes opened the discussion. In his view, the Star was due to a triple conjunction of the planets Jupiter and Saturn in BC 7 (timewise, just within our acceptable limits). Because the planets move round the Sun at different distances and in different periods, there are times when they seem to lie close together in the sky, and these line-of-sight effects can be impressive. For example, on the evening of 6 April 2000 Saturn was fairly high in the sky, with the much more brilliant Jupiter some way below, and Mars just to the right of Jupiter; there too was the thin crescent Moon – a beautiful display. Between May and December of BC 7 Jupiter and Saturn bypassed each other three times, and it must have been interesting to watch, even though Saturn is never really brilliant, and the two planets were never sufficiently close in the sky to merge into a single starlike body. Neither would the planets move in the way described by St Matthew, and the triple conjunction was a prolonged affair; the Wise Men would have known about it – and if Herod had wanted to see it, he would simply have had to go outdoors and look.

Mark Kidger followed with a rather more elaborate theory involving several phenomena. First, the triple conjunction of Jupiter and Saturn in BC 7; then a period in February of the following year when

▶ **The Star of Bethlehem.** In 1304 the great painter Giotto produced his picture The Adoration of the Magi, which showed Halley's Comet as the Star of Bethlehem – though we now know that the comet was seen at least 10 years earlier!

Jupiter, Saturn and Mars were close together in the constellation Pisces, the fishes – the same region as that of the triple conjunction; more planetary conjunctions in BC 5; and then also in BC 5, the outburst of a nova. A nova is due to the flare-up of a faint star, lasting for a few days, weeks or months before subsiding. Note that a nova is very different from a supernova; it is much less violent, and at the end of the outburst the star returns to its old state, though it is true that different novae behave in different ways. There are rather vague Chinese reports of a nova seen around this time, and Mark Kidger associates it with a now-obscure variable star known as DO Aquilae, though he is the first to admit that the identification is very uncertain. In this theory, the nova was simply the culmination of a number of unusual events in the sky, indicating to the Wise Men that the birth of the Messiah was imminent. But again, all the phenomena would have been widely seen – and how does one account for the movement described by St Matthew?

My own idea is that the Star may have been two or more meteors. As a comet moves through space, it leaves a dusty trail in its wake. When the Earth passes through such a trail, small particles dash into the upper air, become heated by friction against the air particles, and burn away to produce the luminous streaks which we call shooting stars. What we see, of course, is not the particle itself, which is of sand-grain size, but the effects in the atmosphere due to its suicidal plunge. Meteors burn away at around 60 km (40 miles) above sea-level, and are quite harmless. (Larger bodies which land more or less intact and can even blast out craters are called meteorites; they are not associated with comets.)

Can we assume that the Wise Men saw a brilliant meteor, rising in the east, crossing the sky in a westward direction, lasting only a few seconds, but leaving a trail that might persist for hours? And is it possible that later another meteor was seen moving in the same direct line? We can take matters a step further. At 9.05 pm on 9 February

◄ **The Great Meteor of 7 October 1868**, *in a painting by an unknown artist. The meteor was so brilliant that it attracted widespread attention, lasting for several seconds and leaving a trail that persisted for minutes. Could a meteor have been the Star of Bethlehem?*

1913, observers in Canada saw a brilliant meteor which moved across the sky with 'peculiar, majestic, dignified deliberation'. No sooner had it disappeared than others followed one after the other, all moving at the same deliberate pace, leaving trails streaming behind them. This is the best authenticated case of a 'meteor procession', and anything comparable would certainly have taken the Wise Men by surprise. Nobody well away from that part of the desert would have seen anything at all. At least meteors do show obvious motion.

Well – planetary conjunctions, novae, meteors, or something quite different? We did not pretend to give any real answers, and we agreed to differ, but it made for an entertaining and lively discussion. I suppose there is always a chance that new evidence will be found, but after two thousand years it does not seem very probable, and all in all I think we must resign ourselves to the fact that the mystery of the Star of Bethlehem will never be solved.

3 • DRY VALLEYS AND MARTIAN MICROBES

The planet Mars has always had a special fascination for us. It is less unlike the Earth than any of the other worlds in the Solar System, and has been regarded as suitable for life. But what are the real prospects for life on Mars, intelligent or otherwise?

Superficially, the situation does not look unpromising. Surrounding the Sun there is a zone, known as the Habitable Zone or Ecosphere, where the temperature conditions are tolerable. Venus is on the inner edge of the zone, Earth comfortable in the middle, and Mars very close to the extreme outer edge. If Mars had the same mass as the Earth, and the same sort of atmosphere, conditions there would be tolerable temperature-wise. In fact, Mars is much less massive than our own world; the escape velocity is only just over 5 km (3 miles) per second, and the tenuous atmosphere is made of chiefly carbon dioxide. Surface seas or even lakes are absent, but at least there is plenty of ice at the poles.

A century ago it was still widely believed that there was intelligent life there, and that the Martians – whatever they might be – had built

a planet-wide irrigation system to carry water from the icy poles to the warmer deserts close to the equator. Percival Lowell, a gifted if obstinate American, set up an observatory at Flagstaff in Arizona, equipped it with a fine 24-inch refractor, and drew up an elaborate network of canals on Mars: 'That Mars is inhabited by beings of some sort or another is as certain as it is uncertain what these beings may be.' Sadly, there are no canals; they were due to nothing more than tricks of the eye. I remember my first view of Mars with Lowell's telescope, soon after World War 2, when I was one of NASA's Moon-mapping team. Would I see canals? I did not, and I am delighted that I failed!

But disposing of Lowell's Martians does not necessarily mean that the planet is sterile. There could be simple life-forms, probably below the surface. Several spacecraft have made controlled landings there, and carried out on-the-spot tests; nothing definite has been found, but the question is still very open. Surprisingly, important clues may come from Antarctica.

It has been claimed that certain meteorites have been blasted away from Mars, and have subsequently landed on the Earth. One of these, known by its catalogue number of ALH 84001, was found in the Allan Hills of Antarctica and was said to contain traces of tiny Martian organisms. I am highly sceptical. The meteorite may be Martian, though even about this I have my doubts, but the 'organisms' may not be indicative of life, and in any case are almost certainly due to terrestrial contamination. We must wait for authentic samples of Mars, brought here by a sample-and-return rocket. This should be possible in the reasonably near future, but for the moment we must wait and see.

◄ **Mars at opposition in 1995,** in an image taken by the Wide Field Planetary Camera 2 on the Hubble Space Telescope. Dark areas, once thought to be vegetation, are now known to be regions of coarse sand.

▶ **Glaciers shaped Mars' surface** as recently as a few million years ago. In this Mars Express image, an hourglass shape has been produced by a glacier flowing downhill from a smaller into a larger crater.

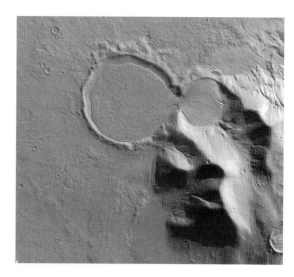

◀ *Mars from the Hubble Space Telescope.* The image was taken on 27 June 1997 to check on weather conditions near the site where Pathfinder was due to land on 4 July.

If there is life on Mars, what will be its nearest Earth equivalent? This brings us on to Antarctica's 'Dry Valleys'. In these valleys there is practically no water, freezing winds blow along them, and for two months each year the Sun remains below the horizon. Conditions there are as hostile as anywhere on the globe. Indeed, allowing for the denser atmosphere, it is fair to say that the conditions are positively 'Martian'. Yet there is life there; there are bacteria in abundance, thriving despite the unfriendliness of the terrain.

In past ages Mars almost certainly had oceans, and we can draw up a rough timetable:

From 4.2 to 4.8 thousand million years ago: plenty of surface water, with oceans and rivers.

3.8 to 3.1 thousand million years ago: water restricted to covered lakes, increasingly saline.

3.1 to 1.5 thousand million years ago: liquid water still to be found in porous rocks and sediments.

1.5 thousand million years ago to the present time: no surface water, ice in permafrost, water surviving only at appreciable depths below the surface.

Now compare Mars with the Dry Valleys of Antarctica, where microbes are plentiful. The most evident of these microbes are what we call 'cyanobacteria', which are found in Antarctica wherever water is available, no matter how briefly. They are also found in the fossil record of the Earth dating back at least $2\frac{1}{2}$ thousand million years ago. Microbes actually live inside rocks – notably the translucent sandstone right at the edge of the Antarctic Polar Plateau. They are being studied by a technique known as Raman Spectroscopy, developed in 1928 by C.V. Raman. The same technique can be used on Mars – when we manage to set up equipment there. It would not be at all surprising to find the same types of organisms. The first Raman

Spectrometer will be flown to Mars by 2007, though it would be optimistic to expect a manned mission much before 2020.

If life appeared on Mars, it could well have led to the evolution of what are termed photosynthetic microbes, similar to those of the Dry Valleys. Pigments would be developed, both to harness the energy of sunlight for photosynthesis (the procedure of removing carbon dioxide from the atmosphere and replacing it with oxygen) and to protect the cells from damage by ultraviolet radiation from the Sun. Some of these pigments are familiar, such as carotene (Vitamin A), which gives carrots their orange colour. I am not suggesting that there are carrot crops on Mars, but Vitamin A is a distinct possibility. The Raman Spectrometer to be flown to Mars will first be used to test the samples from the Dry Valleys – collected mainly from a British desert research site, piquantly known as Mars Oasis. (It was named from the nearby Mars Glacier; the name was given long ago!)

Detailed pictures of Mars have been obtained from recent spacecraft, notably the Mars Global Surveyor, which has been orbiting the planet ever since the autumn of 1997 and is still (2005) working excellently. There is a strong resemblance between, say, the fluvial delta produced in Gusev Crater by an ancient Martian river in

▲ **Plates on Mars.** The Mars Express orbiter photographed these unusual plates. Thought to be blocks of ice floating on a dust-covered frozen sea, they bear a resemblance to ice blocks off Earth's Antarctica.

Ma'adim Vallis, and an ice-covered hypersaline lake, Lake Vanda in Antarctica's Wright Valley. Note that microbes can survive in hypersaline environments in Antarctica down to a temperature of –50°C. Cyanobacteria form 'mats' at the bottom of Antarctic lakes such as Don Juan Pond; these are termed stromatolites. Stromatolites have been seen in existence for around $3\frac{1}{2}$ thousand million years. Can there be stromatolites on Mars? There seems to be no reason why not.

Long ago, when there were Martian Oceans and Lakes, there was probably life on the surface. If there is any surface life now it must be very lowly, but there could be swarms of living things in a deep, dark biosphere where sunlight never penetrates. There could well be a vast ocean.

Time will tell. Certainly the exploration of Mars will go on as quickly as possible, and nobody can doubt that the results will be exciting. And well before January 2102, the whole planet may have been transformed, the new 'Martians' will have arrived.

4 · GALACTIC WHIRLPOOLS

Most people believe that when you unplug a bath full of water the ensuing 'whirlpool' will rotate anticlockwise in the northern hemisphere and clockwise in the southern. There is a grain of truth in this, because we live on an Earth which is rotating on its axis, though in practice this so-called Coriolis Effect is masked by small random movements in the water. On a larger scale, with tropical storms, the Coriolis Effect does become dominant.

Spiral structures in space are common – even some planetary nebulae, such as the Helix Nebula in Aquarius, are spiral. (A planetary nebula is misnamed; it is simply a very evolved star which has thrown off its outer layers, so that it is not truly a nebula and has absolutely nothing to do with a planet.) But these spiral forms are not due to Coriolis forces, and neither are the beautiful structures once called spiral nebulae but now known to be independent galaxies, some of them similar to our own Milky Way system.

The first galaxy seen to be spiral was M51, the 'Whirlpool', in the constellation of Canes Venatici, the Hunting Dogs. The observation was made by the third Earl of Rosse in 1845, with the strange but highly effective 72-inch reflector which he had set up at Birr Castle, in Ireland. (The telescope is now fully operational

▼ **The spiral galaxy M83,** as imaged by the Kitt Peak 4-metre telescope. It was observed by Lord Rosse using the great Birr Castle reflector.

again, for the first time since 1909; I am delighted to say that I had a good deal to do with its restoration!) Lord Rosse also saw other spirals such as M99, M100 and the lovely M83 in Hydra. M83 was also drawn by William Lassell in 1860, using his 48-inch reflector at Malta. It is unusual inasmuch as it seems to be intermediate in type between a normal spiral and a barred spiral galaxy, where the arms issue from the ends

▲ *The spiral galaxy* **M33,** *also known as the Triangulum Galaxy. The reddish areas are regions of star formation, filled with hot hydrogen gas.*

of a sort of bar passing through the centre of the system and along the main plane.

What seems to happen is that a pressure wave propagates outward from the nucleus of the galaxy. This wave triggers star formation as it passes through the disk material, and spiral arms develop. The causes of the pressure waves are less clear. Some may be due to explosions in the centre of the system, possibly as material falls into a massive black hole – or large numbers of supernovae. But there is little doubt that pressure waves develop when galaxies collide, and we know that this does happen. Direct collisions between stars must be vanishingly rare, but the material between the stars will collide throughout the encounter –and the effects will be very pronounced. For example, M51 has a companion galaxy – which is itself spiral. The two are joined by a bridge of material which has been pulled out by tidal forces. The smaller system, NGC 5195, passed by M51 about 200

▼ **M51, or the Whirlpool Galaxy,** is seen here with its smaller companion galaxy,

NGC 5195. The bridge of material that joins the two galaxies can be clearly seen.

million years ago. An even better case is given by the Antennae galaxies in Corvus, NGC 4038 and 4039 (Caldwell 60 and 61). The collision was almost head-on, and images taken with powerful telescopes show the two antennae very clearly.

Our own galaxy will collide with M31, the Andromeda, in about 3000 million years. The two spirals will be destroyed as such, and the end product is likely to be a single large elliptical system. Many stars will be thrown off into intergalactic space, and our Sun could well be one of those expelled stars. Luckily 3000 million years is a very long time ahead, and at present the Andromeda galaxy is over 2,000,000 light-years away, so that there is no immediate cause for alarm.

By no means all galaxies are spiral, some are elliptical, some are spherical and others quite formless. There is plenty of variety, but one cannot doubt that these graceful 'whirlpools' are among the most beautiful objects in the universe.

5 · THE GREAT BEAR

O f all the constellations, the two which are best known are the
Great Bear and the Southern Cross. From Britain, the Cross is
never to be seen, but the Bear never sets, and there can be few people who cannot recognize it.

In fact its stars are not particularly brilliant, and cannot rival the
splendour of Orion or the Scorpion, but there is a very characteristic
pattern, known popularly as the Plough or, in America, the Big
Dipper. It consists of seven stars. They are:

Star	Greek Letter	Magnitude	Distance, l.y.	Luminosity, Sun=1	Spectrum
Dubhe	Alpha	1.79	75	60	K0
Merak	Beta	2.37	62	28	A1
Phad	Gamma	2.44	75	50	A0
Megrez	Delta	3.31	65	17	A3
Alioth	Epsilon	1.77	62	60	A0
Mizar	Zeta	2.09	59	50+11	A2+A6
Alkaid	Eta	1.86	108	450	B3

Obviously the stars are not all alike. Dubhe is orange, the rest
white or bluish; Alkaid is the most powerful, while Megrez is the
faintest of the seven by a magnitude. Neither are they genuinely
associated. Five of them do make up a moving cluster, and are trav-
elling through space in the same direction at much the same rate,
but Dubhe and Alkaid are moving in the opposite direction, so that
after a sufficiently long period of time the Plough pattern will
become distorted. However, the proper motions are too slight to be
noticed even over periods of many lifetimes. King Canute, Julius
Caesar and the builders of the Pyramids would have seen the
Plough just as we do today.

The relative faintness of Megrez stands out at once. There have
been suggestions that in ancient times it was the equal of the rest,
but this seems decidedly improbable. Megrez is a perfectly normal,
stable, main sequence star, and would not be expected to show sec-
ular variation.

(It is always unwise to trust the old estimates too far. Other cases
are on the file: for example, it is claimed that Castor was once
brighter than Pollux – but all these alleged changes are highly suspect.)

Mizar is the most celebrated star in the Plough. Close beside is a
fainter star, Alcor, easily visible with the naked eye, and telescopically. Mizar itself is seen to be double, with one component appreciably
brighter than the other. The two Mizars make up a binary system, but
the revolution period is very long indeed; the bright component is

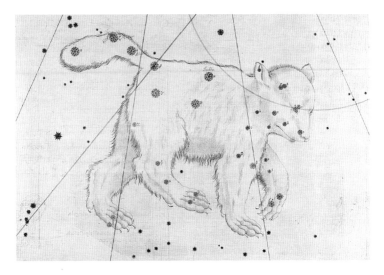

▲ **Ursa Major,** or the
Great Bear, seen here as it
appears in Johann Bayer's
Uranometria of 1603.

◄ **The seven main stars
of Ursa Major,** the Great
Bear. They are, from the
left, Alkaid; Mizar, with
Alcor; Alioth; Phad and
Megrez; Merak and
Dubhe (the Pointers).

itself a spectroscopic binary. Alcor is a true member of the system
inasmuch as it has the same motion through space, but the revolution period is of the order of millions of years, because the distance
between Mizar and Alcor is several light-years.

Between Alcor and the Mizars is an 8th-magnitude star, named
Sidus Ludovicianum in 1723 by the courtiers of the Emperor
Ludwig V, who believed that it had appeared suddenly. It is visible
with binoculars but not with the naked eye, and there is a minor
mystery here. The Arabs of a thousand years ago said that Alcor was
difficult to see except on a very clear night, but there is nothing elusive about it now. Bearing in mind that the Arabs were keen-sight-
ed, and did not have to cope with light pollution, could they have

been referring to Ludwig's Star? Again this seems unlikely; the star shows no sign of variability. Incidentally, it is not a member of the Mizar group, and lies well away in the background.

Any casual glance will show that Dubhe is orange, contrasting with the bluish-white of the other Plough stars. It is a close binary; the companion is not faint (magnitude 4.8), but the separation between the two is never as much as one second of arc. The period is 45 years. Merak and Dubhe are known as the Pointers, because they show the way to the pole star in Ursa Minor (the Little Bear). The pole star, Polaris, is of the 2nd magnitude and is very slightly variable; it is 6000 times as luminous as the Sun, so that it is far more powerful than any of the seven stars of the Plough. Ursa Minor itself is a faint constellation, in shape like a dim and distorted version of the Plough. Apart from Polaris, the only brightish star is the orange Kocab, also of the 2nd magnitude. Between Kocab and Alkaid lies Thuban (Alpha Draconis), which was the pole star at the time when the Egyptian Pyramids were built. Twelve thousand years hence, the north pole star will be the brilliant Vega, in Lyra.

The Plough is only part of the large constellation of Ursa Major. There is a legend about it. Callisto, daughter of King Lycaon of Arcadia, was so beautiful that her loveliness exceeded that of Juno, queen of Olympus. This was not to Juno's liking, and ill-naturedly she turned Callisto into a bear. Years later Callisto's son Arcas went hunting, he encountered the bear, and was about the kill it when Jupiter intervened, catching both bears by their tails and springing them up into the safety of the sky. That is why both Ursa Major and Ursa Minor have tails of decidedly un-ursine length!

There are various objects of interest in the constellation. Xi (Alula Australis) is an easy binary, with two slightly unequal components

▶ *The Owl Nebula, M97,* does indeed resemble an owl's face when properly seen. It is faint, however, and can be hard to locate.

▲ **M109 in Ursa Major** *is a barred spiral galaxy, seen here in an image taken* *by the 0.9 m telescope at Kitt Peak. It is in the same group of galaxies as M108.*

and a separation of 12.6 seconds of arc; the orbital period is 60 years and this was the first binary to have its orbit worked out – by Felix Savary in 1830. At a distance of 25 light-years, Xi is one of our nearer stellar neighbours. Close beside it is the obviously orange Nu (Alula Borealis), 150 light-years away and over 100 times as powerful as the Sun.

Z Ursae Majoris, not far from Megrez, is a red semi-regular variable, easy to locate with binoculars; the range is from magnitude 6.8 to 9.1, and there is a rough period of 196 days. There are a few Mira variables, but all remain below naked-eye range.

There are two interesting galaxies, the spiral M81 and the irregular M82, which are around 8,500,000 light-years away, and have marked effects on each other; M82 is a starburst galaxy, and a strong radio source, while M81 is a conventional spiral. Both are easy telescopic objects, and M81 is detectable with good binoculars. M108, near Merak, is a much dimmer and looser spiral, more or less edgewise-on to us.

In the same telescopic field with M108 is M97, the 'Owl Nebula' – a planetary, discovered by Messier himself in 1781. This is one of the most elusive of the Messier objects, and is difficult to find with a small telescope; it is well beyond binocular range. The central star is only of the 14th magnitude.

Ursa Major is rich in faint galaxies, but lacking in gaseous nebulae. The celebrated Whirlpool Galaxy, M51, is not far from Alkaid, but lies over the border of the adjacent constellation of Canes Venatici (the Hunting Dogs).

Look for the star 47 Ursae Majoris, near the little triangle made up of Psi, Lambda and Mu (note that Lambda is white, Mu very red). 47 is of the 5th magnitude, and is 42 light-years away. It has been found to be attended by a planet, moving round the star at a range of 200 million miles in a period of 1095 days; as the mass is 2.4 times that of Jupiter, it must be a hot gas-giant. There is also evidence of a third body in the system, farther out and less massive than Jupiter. For all we know, there may also be a planet similar to the Earth.

There is, then, plenty to see in Ursa Major. It is a permanent feature of the night sky as seen from Britain; from far-southern countries such as the Southern Island of New Zealand it is out of view, and the sky seems strange without it.

6 · MOONSHINE?

In 1968 Astronauts Lovell, Borman and Anders, in Apollo 8, became the first men to go round the Moon in a spacecraft. In the following year first Neil Armstrong, then Buzz Aldrin stepped out on to the bleak rocks of the lunar Sea of Tranquillity. Since then there have been five more successful trips and one partial failure, rocks have been brought back, scientific equipment left on the Moon, and vast amounts of data collected. And yet – *mirabile dictu!* – there are still people who believe that the entire Apollo programme was faked by NASA conspirators.

As soon as I hear the term 'conspiracy theory' I know that I am dealing with cranks. But the idea of a faked Apollo has become so widespread that it must be disposed of, particularly since there are folk who sincerely believe it – just as there are some people who stubbornly maintain that the Earth is flat.

The idea came from a science fiction film, *Capricorn One*, in which NASA sportingly collaborated; this dealt with a faked Mars landing, and it was very cleverly shot as well as being a good story. But I am sure that neither NASA nor anyone else expected any repercussions, and what actually happened came as a complete surprise. Moreover, the arguments put forward by the 'conspiracy group' sounded outwardly convincing!

In fact, it is always easy to bolster a crazy theory with plausible arguments. The flat Earth idea is a case in point. How do you prove that the Earth is a globe? By watching a ship pass over the horizon, when the hull disappears first and the upper part remains in view? In theory this is positive proof, but how many people have actually seen it? Very few, and moreover, refraction effects can at times lift the entire ship above sea horizon. In an aircraft, fly on a straight and level course and your height above the ground will remain constant; if the Earth were a globe, your height would increase. And turning to religion, how could angels stand at the four corners of a round Earth?

You can see how easy it is! So let us look at some of the arguments of the 'conspirators'.

(1) The rocks, allegedly brought back from the Moon, have been manufactured on the Earth. – Almost 400 kg (about 850 lbs) of rocks were brought back by the Apollo astronauts (incidentally, far more than could have been carried by the unmanned Russian landers). The differences between lunar and terrestrial rocks are very marked. The rocks from the Moon were obviously formed in an environment lacking oxygen and water; there are isotopes resulting from bombardment by cosmic rays and the solar wind, from which we on the Earth are protected. There is also extensive pitting from micrometeorite impacts, and there are minerals which simply do not occur in our rocks. One eminent geologist commented that it might just about be possible to fabricate 'lunar' rocks, but it would take so long and be

so costly that it would probably be easier to go to the Moon and bring back genuine specimens!

(2) Intense radiation found in space would have killed the astronauts even before they reached the Moon. – True, the Van Allen belts can be dangerous, and they dip down toward the Earth's surface over the Atlantic (the 'South Atlantic Anomaly'). But men have remained on spacecraft such as Mir for long periods, and have returned safely. Consider, too, the effects on photographic film. Holiday-makers who have been unaware of the problems caused by airport X-rays have often lost their cherished pictures; but the films brought back from the Moon show no sign of fogging.

(3) Neil Armstrong was shown coming out of the spacecraft and on to the lunar surface, but how can these pictures have been obtained? – By a camera on the Apollo lunar module, switched on automatically.

(4) The lunar surface is dust-covered, yet no dust was disturbed as the lunar module came down. – It was. The movie record of the landing sequence clearly shows dust being blown around by the descent engine. With a dust layer of only an inch or two deep, this is just what would be expected.

(5) The lunar sky is black all the time, because to all intents and purposes there is no atmosphere; yet no stars are seen in the photographs

▲ **Neil Armstrong,**
descending from Apollo 11
to the Moon's surface. The
photograph was taken by
a camera on the lunar
module.

taken from the surface. – Of course not. The astronauts were using ISO 150 colour film, with typical exposures of the order of 1/25 of a second at f/8 or f/11. To record stars on such a film would mean an exposure of more than 20 seconds at f/2.8. If you doubt this, try for yourself on the next dark night. (Incidentally, one of the first things I asked Neil and Buzz on their return was whether they could see stars in the lunar daytime. The answer was 'yes' – but only by shielding their eyes against the glare of the rocks, and dark-adapting for a brief period.)

(6) Some of the shadows cast both by rocks and the spacecraft point in different directions. – This is natural. Shadows only point in the same direction when they fall upon a completely flat surface – and this is something that you do not find on the Moon. A simple demonstration can explain this. Set up three posts, and photograph their shadows thrown on to an undulating surface. The results speak for themselves.

(7) Lunar shadows are black, so no detail should be seen inside them; the only source of light is the Sun, which more or less ranks as a point source. – The Sun is not the only source of light; there are bright rocks all around. If the pictures had been taken in a film set, with multiple light sources, there would have been multiple shadows. A good example of this is provided by a floodlit football match.

(8) Various pieces of equipment were left on the Moon, but these have never been shown on any photograph taken subsequently from

◄ **Neil Armstrong,** during the first extra-vehicular activity (EVA). The horizontal bar supporting the flag is clearly visible.

► **Buzz Aldrin on the Moon,** photographed by Neil Armstrong, whose reflection can be seen in Aldrin's visor.

Earth. – They are simply too small, as any elementary calculation will show.

(9) If the astronauts had their cameras at chest height, it would have been impossible for one astronaut to have photographed another astronaut's head. – Not if they stood at different levels; and in any case, the cameras could be tilted.

(10) The published photographs are too good to have been taken by untrained photographers. – The astronauts had been very well trained, and they had the best possible equipment. But, naturally, only the good images were released; there were also many failures.

(11) The American flags fly horizontally, and do not flap. – There is no wind to make them flap, and in any case they are supported by a horizontal bar.

(12) The intended landing site for Apollo 13 was Fra Mauro, but this was in darkness. – Only when Apollo 13 was launched. When it should have landed, the Sun had risen over Fra Mauro. Of course, Apollo 14 did land there.

(13) The videos of the lunar rovers, showing the astronauts of Apollos 15, 16 and 17 driving around, were simply half-speed replays

that had been filmed on Earth. – Not so! The whole effect would be different. And look at the path of the dust thrown up by the rover's wheels; it describes a perfect parabolic path, and this can only happen in a vacuum. On Earth, the atmosphere would make the dust particles fall to the ground more quickly.

The other claims put forward by the conspiracy-theory believers are even less convincing – and in any case, can one seriously maintain that a hoax on this scale would have been kept secret for so many years? Before long, people will go back to the Moon, and there they will find the equipment left behind by the Apollo pioneers. No doubt they will collect it, and move it to a lunar museum.

Of course, nothing will make the conspiracy-theory followers change their minds. It takes all sorts to make a world, or for that matter two worlds, and all we can really say about these folk is that if ignorance is bliss, they must be very happy!

— 7 · FORTY-FIFTH ANNIVERSARY —

The first *The Sky At Night* programme was transmitted in April 1957. Since then a great deal has happened, and astronomy has been completely transformed. In 1957 the Space Age lay in the future; the great radio telescope at Jodrell Bank was barely completed; among optical telescopes the 200-inch Hale reflector was in a class of its own; objects such as pulsars and quasars were not only unknown, but unsuspected. In our Solar System, it was widely believed that there were oceans on Venus, and extensive vegetation tracts on the surface of Mars.

We know better today, and for our 45th anniversary programme we decided to draw some comparisons between 'then' and 'now'.

Venus is a case in point. It is about the same size as the Earth, and closer to the Sun than we are (on average 180,000,000 km [67,000,000 miles], as against our 150,000,000 km [93,000,000 miles]). Its surface is permanently hidden by its dense, cloud-laden atmosphere, and in 1957 even its axial rotation period was unknown.

▼ **Venus from Magellan.** *Launched in 1989, the Magellan probe transformed* *our knowledge of the planet. Its radar mapped 98 per cent of the surface.*

Carbon dioxide acts in the manner of a greenhouse, and blankets in the Sun's heat, so that Venus had to be a hot place, but the idea of oceans did not seem at all extravagant, and was supported by eminent astronomers such as D.F. Menzel and F.L. Whipple. If so, then the atmospheric carbon dioxide would have fouled the water, producing oceans of soda-water. Life on Earth probably began in our warm seas, and the same might have been true for Venus. It was only after 1962, when the first successful Venus mission flew past the planet, that the attractive 'marine theory' had to be abandoned. Alas, the conditions there are hopelessly hostile. The whole surface has been mapped, largely by the US Magellan probe of the 1990s; there are lowlands, uplands, craters and volcanoes, with lava-flows everywhere. In the foreseeable future, the chances of a manned mission to Venus seem to be effectively nil.

We were also wide of the mark with regard to Mars. The dark areas were tacitly assumed to be vegetation, probably old sea-beds; we knew there were no Martians, but the planet was generally believed

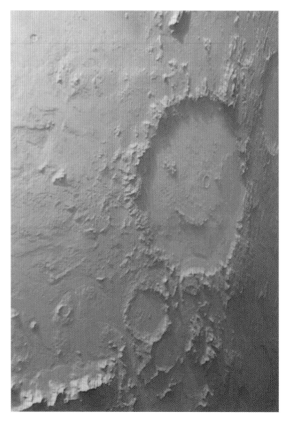

◀ *Mars from Mars Global Surveyor. The crater Galle, nicknamed the 'Happy Smiling Face', is 230 km (140 miles) across. Such features were unknown before spacecraft visited the planet.*

▶ *Quasar 3C273,*
imaged by the 4-m Mayall
telescope at Kitt Peak. It
radiates a hundred times
more light than the
brightest ordinary galaxy.

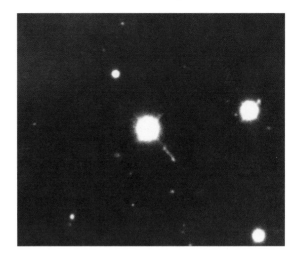

to be lacking in mountains and craters, while some astronomers still maintained that the canals had a 'basis of reality'. It took the space-craft to reveal the vast, massive volcanoes such as Olympus Mons, which towers to a height of 24 km (15 miles) above the surrounding terrain and is topped by a 64-km (40-mile) caldera; the surface is heavily cratered, so that Mars has been bombarded in the remote past. There is little doubt that there were once seas; today the water has gone from the surface, though there may be liquid not far below the crust. The Martian climates change over long periods, because the tilt of the axis ranges between 35 degrees and only 14 degrees to the perpendicular to the orbital plane; this does not happen with the Earth – thanks to our Moon, which keeps the tilt value very close to its present $23\frac{1}{2}$ degrees.

Of course, in 1957, our best views of the planets were obtained with Earth-based telescopes and photographic plates. We were bare-ly at the start of the 'electronic revolution', and computers of the 1950s seem primitive now. Space missions have shown us what the planets and their satellites are really like; we have seen the ice cliffs of Miranda, the wildly active volcanoes of Io, and the chilly nitrogen geysers of Triton. The Hubble Space Telescope has even revealed a certain amount of detail on Pluto, which remains the only planet not to have been by-passed by at least one probe.

In 1957 there was still keen rivalry between the 'big bang' theory of the universe and the 'steady-state' theory, according to which the uni-verse has always existed and will exist for ever. Observational devel-opments led to the demise of the steady-state hypothesis, but we have to admit that we are still at a loss with regard to the creation itself. It is all very well to say that space, time and matter came into existence simultaneously, between 13,000 million and 14,000 million years ago (the accepted value as of May 2002) but *how*? We simply do not know, and in this respect we are no better off than we were in 1957.

Quasars came upon the scene in the 1960s, when it was found that some dim objects which looked like blue stars were in fact immensely remote and almost incredibly luminous. They have proved to be the nuclei of very active galaxies, and to be racing away from us at well over 90 per cent of the velocity of light. Great new telescopes can allow us to explore the far reaches of the universe, but we have not yet come to 'the outer limits'. Perhaps we will do so before *The Sky at Night* celebrates its 50th anniversary!

Pulsars caused a stir when they were discovered by Jocelyn Bell-Burnell, who joined me in one of our programmes to describe what happened. For a brief period it was even thought that these strange, 'ticking' pulses might be artificial. Then there are black holes, which are surely the most bizarre objects of all – cut off from the rest of the universe. They too were unknown in 1957.

I suppose that one of the greatest advances has been the revelation that other stars have planets of their own. Proof came only in the 1990s, and even now (May 2002) we have never had an actual view of an extra-solar planet, though there have been several false alarms. If other Earths exist, why not other civilizations? It seems logical, and now we are sure that planetary systems are common, there is no reason to regard ourselves as unique or even unusual. One day, contact must be made, unless all our current ideas are completely wrong.

During our first 45 years we have taken *The Sky at Night* to many parts of the globe. We have been to the great observatories in Chile, Hawaii and Australia; we have been down a gold-mine in the Black Hills to hunt neutrinos sent out by the Sun; we have chased solar eclipses in the Caribbean and atop a mountain in what was then Yugoslavia. We have been joined by many of the world's leading astronomers, some of whom, alas, are no longer with us. We have been to the rocket bases in Florida and Russia, and we have been at NASA Headquarters during the missions to the outer planets. It has all been tremendously enjoyable – and, I hope, useful too.

There has been one interesting coincidence. Our very first programme was ushered in by a comet, Arend–Roland, which was a bright naked-eye object for a week or two in late April 1957. For our 45th anniversary, we were again provided with a naked-eye comet, Ikeya–Zhang. But there the similarity ends. Ikeya–Zhang will be with us again in 341 years' time. Arend–Roland will never return; an encounter with Jupiter has thrown it out of the Solar System, and condemned it to wander between the stars. I will always have great affection for it. I wonder where it is now?

8 · SOUTHERN EYES

Before the 20th century, most astronomy came from the northern hemisphere. Nowadays the emphasis has largely shifted southward. There are two reasons for this. First, the southern skies are less affected by light pollution; secondly, the southern stars are of particular interest. We Europeans never see the Clouds of Magellan, the brightest parts of the Milky Way, or the Southern Cross.

South Africa led the way, and then came Australia. Today the main Australian observatory is at Siding Spring in New South Wales, well north of Sydney and not far from the little town of Coonabarabran. Australia has no high mountain, but conditions at Siding Spring are as good as they are anywhere on the continent, and *The Sky at Night* has been there several times.

▼ **The dome of the United Kingdom Schmidt telescope**. *An outstation of the Royal Observatory, Edinburgh, from 1973 to 1988, the* UKS *is now part of the Anglo-Australian Observatory. During its first 20 years of operation it conducted a photographic survey of the southern sky.*

The two main instruments are the AAT or Anglo-Australian Telescope, with mirror 153 inches across, and the 49-inch UKS or United Kingdom Schmidt (respectively 3.9 metres and 1.2 metres aperture, if you prefer metric).

The AAT is a superb instrument; it has been used for all branches of research, and with it David Malin has taken spectacular photographs which are arguably the best ever produced from an Earth-based telescope. The 1980s saw a new era of statistical studies of stars and galaxies, and fibre optics techniques have been used on the AAT. In late 1997 a fibre-optics robot, the 2dF, was attached to the AAT and began collecting and analyzing on a large scale. (2dF stands for 'two degree field'.) Over a mere five nights, almost 4000 galaxies and 1000 quasars were observed and analyzed – the first time that so many objects have been surveyed in so short a time; previously it would have taken years to collect and analyze those amounts of data.

◀ *The United Kingdom Schmidt is a specialized photographic telescope. Its optics are of the highest standard, and its wide field of view covers 40 square degrees.*

▲ *Parkes Radio Telescope,* in central New South Wales. It forms part of the Australia Telescope *National Facility, a project of major importance to radio astronomy research in the southern hemisphere.*

The 2dF has now completed a major survey of a quarter of a million galaxies, showing the positions of 22,000 quasars out to more than 13,000 million light-years. It has also shown that neutrinos cannot be responsible for all the dark matter known to exist in the universe, though we have to admit that the precise nature of dark matter remains a mystery.

The UKS began work in 1973, mainly to construct a photographic survey of the entire southern sky. To date over 17,000 plates have been taken, providing a huge source of data for astronomers all over the world. For this type of work a Schmidt telescope is more or less essential, because it can cover very wide fields with a single exposure. It is now being used with a new fibre-optics robot, the 6dF, a direct analogue of the 2dF system in the AAT; it can provide redshift measurements for the spectra of very faint galaxies.

Telescope time is always at a premium, and it is significant that during the first year of the 6dF it was allotted 75 per cent of the total time on the UKS. During the current survey, about 90,000 redshift determinations will be made, over an area of 15,000 square degrees. It will take between 2 and $2\frac{1}{2}$ years, and should lead on to important new developments in our studies of the kinematics and perhaps the chemistry of our Galaxy.

One important point should be made here. By modern standards the 153-inch AAT is not a giant telescope; compare it with the VLT or Very Large Telescope in Northern Chile, which has four 8-metre mirrors working together – and the Schmidt telescope at Tautenberg in Germany is larger than the UKS. Yet the pioneer working at Siding Spring has shown that modest telescopes can be used for vital advances in research; everything depends upon the skill of the astronomers who make and use them!

— 9 • BEGINNINGS AND ENDINGS —

It is probably true to say that the 'modern' age of astronomy began in 1687, with the publication of the *Principia* by Isaac Newton. Newton was then Lucasian Professor of Mathematics at Cambridge University. Today, his mantle has fallen upon the shoulders of Professor Stephen Hawking, who currently occupies the Lucasian Chair.

Stephen Hawking was born on 8 January 1942, three hundred years after Newton died. Despite his physical handicap he is widely recognized as the world's leading mathematician, and his contributions have been enormous. In particular, he is looking back to the very early days of the universe.

According to the Big Bang theory, the universe began in a state of high temperature and density, and has been expanding ever since. The universe started off with just energy, and this energy was converted into matter and antimatter. Particles and anti-particles collided, destroying each other and leaving radiation. For some reason or another there was more matter than antimatter at the end of the

▲ *The galaxy cluster* **Abell 2218,** *imaged by the Hubble Space Telescope. The cluster is so massive that its gravitational field bends light from more distant objects to produce the arcs clearly visible in this image.*

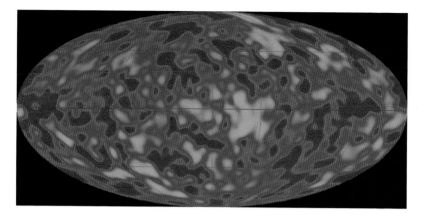

▲ **The Cosmic
Microwave Background,**
*a remnant of the Big Bang.
The tiny variations in*
*temperature, as seen in this
COBE image, are thought to
have triggered the first
galaxy formation.*

formative stage. For the first million years or so the universe was filled with plasma, and was in thermal equilibrium with radiation – a phase known as the primordial fireball.

The remnant of this event was accidentally discovered in 1965 by the American physicists Arno Penzias and Bob Wilson. They found the cosmic background radiation – the signature of the Big Bang. At first they thought that the interference detected at their 'horn antenna' was due to pigeon droppings, but in fact they had discovered the most direct evidence that the universe had passed through a hot, dense phase. This also disproved the 'steady-state' theory, according to which the universe has always existed, and will exist for ever.

General relativity was a new theory of gravity, replacing the earlier ideas of Isaac Newton. Einstein believed that space and time were flexible entities rather than being solid and immutable, and he constructed a complex series of mathematical equations, linking their warping with the matter they contained. According to this theory, objects moving in a gravitational field must follow curved trajectories, simply because space and time are curved. The universe at large is governed entirely by gravity; we can see this as light curves around a galaxy. General relativity was a landmark breakthrough for cosmology. In 1917, before cosmic expansion had been discovered, the universe was generally believed to be static, but relativity predicted that it would either expand or contract. Einstein did not have complete confidence in his own theory, and so he modified it to make it compatible with the static picture of the universe. Had he kept to his original equations, he would have been able to predict the expansion of the universe, which Hubble demonstrated in 1929.

A Catholic priest, Georges Lemaître, realized that if the galaxies were moving apart with time, they must in the past have been much closer together. Keep winding back the clock, and you will reach a time when the galaxies overlapped; go back further still, and you will reach a point where all the matter in the universe was packed into a very small, hot and dense sphere, which is known as the 'primeval atom'.

General relativity works well until you trace things back very close to the Big Bang, when the rules we regard as 'normal' just do not work at all. Our usual laws start working after the first trillion of a trillionth of a trillionth of a second; before that, quantum theory rules. We can use quantum physics to explain how the universe and everything in it – galaxies, stars, planets and life – was born from 'nothing' at the moment of the Big Bang. This is not to say that the universe was created out of empty space; the universe was created out of absolutely nothing, because there was nothing before the Big Bang – not even space and time. Einstein's genius showed him that space and

▼ *Hubble Ultra Deep Field.* The circled galaxies are the remotest known, so that we see them as they were when the universe was very young. Research reveals that they are all small dwarf galaxies.

time are not separate, but are different directions in a four-dimensional spacetime. General relativity defines the curvature of this spacetime and links it with the positions and motions of particles of matter. The curvature bends things towards each other and gives space and time a shape; it has even been described as pear-shaped.

The Hubble telescope allows us to see 90 per cent of the way back to the beginning of the universe, and in microwaves we can look back 99.99 per cent of the way back to the beginning; long before the galaxies were formed we can see the fingerprints of creation. But what can we say about the creation itself?

niverse began as a tiny 'nut', a thousand trillion
cross, filled with a strange substance called vac-
used the universe to expand so rapidly that now,
id million years later, the expansion is still going
nsion go on forever?

s upon the amount of matter in the universe,
e expansion, and whether there is enough vacu-
ed the expansion up. If the density of matter is
al value, the universe will stop expanding, and
tself in a 'Big Crunch' of colliding galaxies.

However, astronomers at the Cerro Tololo
Observatory in Chile have used the rem-
nants of dying stars and supernovae, as
Edwin Hubble did, to measure the expansion of the universe. They find that matter is
not holding the universe together at all. They
expected to find that gravity is slowing the
expansion; instead they found that expansion is speeding up. The galaxies are moving
apart faster than before. This suggests that
there is more vacuum energy, accelerating
the expansion, than there is matter, which
slows the expansion down. If so, then the
universe will presumably expand for ever,
getting colder and emptier; even black holes
will eventually evaporate, dissolving into a
cloud of particles and radiation. The future
of the universe would be long and cold – but
perhaps that is better than short and fiery!

Quantum physics also leads on to some
bizarre possibilities. During its earliest
moments, parts of the universe may have
receded from each other at speeds greater
than that of light, and multiple universes
could have been created simultaneously. This
is certainly difficult to visualize.

Moreover, it seems that over 90 per cent
of the universe is invisible to us. This mysterious missing matter is known as 'dark
matter'. We are sure that it is there, because
of its effects on the movements of stars and

◀ **Spiral galaxy NGC 3370,** from the Hubble Space Telescope. In 1994, a type Ia supernova exploded in NGC 3370.

Measurements of supernovae have allowed astronomers to determine that the expansion of the universe is accelerating.

galaxies, but it cannot be seen directly, because it emits little or no radiation. Cosmologists are very anxious to find out how much dark matter there is, and how it is affecting the expansion of the universe. Unfortunately we do not know just what dark matter is. It could be due to large numbers of brown dwarf stars – that is to say, Jupiter-sized objects which never became hot enough for nuclear reactions to begin. It may be due to strange-sounding things such as WIMPS (Weakly Interacting Massive Particles) or MACHOS (Massive Compact Halo Objects). But the most exciting possibility is that the dark matter is not actually in our universe, but is in another universe – a fifth dimension. In our *Sky at Night* programme, Stephen Hawking made a telling comment:

'The biggest challenge to the human race in the coming century will be to survive to the end of it. We are like children playing with sticks of dynamite; also, something could wipe us out, such as a genetically engineered virus or a runaway greenhouse effect. But if we do make it to the 22nd century, we should begin to move into space. I think we will have to go to Alpha Centauri, or beyond. We already occupy the only decent planet in our Solar System, but once the human race has spread through the Galaxy, it should be safe.'

10 • THE SOUTH DOWNS PLANETARIUM

On 5 April 2002 the Astronomer Royal, Professor Sir Martin Rees, officially opened the South Downs Planetarium. It was a great moment, and the culmination of 7 years of hard work by a small band of very determined and enthusiastic people. With no official help, and total indifference from bodies such as the National Lottery and Ministry of Science, we had started from scratch, raised over £300,000, and brought the Planetarium into being.

Some people still do not know the difference between a planetarium and an observatory. There are no telescopes in a planetarium, which is purely for education and entertainment. A special, very complex projector produces an artificial sky on the inside of a large dome; the effect is amazingly realistic, and it is hard to realize you are not in the open air. 'But,' I can hear you saying, 'why create an artificial sky when the real one is available?'

▲ *The South Downs Planetarium,* in Chichester; with me is John Fletcher. It has proved to be a great success, and is visited by many people.

There are many reasons. First, the real sky is not always available – clouds and fogs see to that, to say nothing of the ever-increasing problem of light pollution (anyone living in a city will be lucky to see any stars at all). Next, the planetarium sky can produce phenomena that the real sky seldom does; for example, would you like to see a total eclipse of the Sun, or a great comet? If you stay in England you will have to wait until 2090 for your next total eclipse, and a brilliant comet may not appear for decades, but in the planetarium sky they can be produced at any moment. Thirdly, the sky can be used for star recognition, and lines such as the meridian can be put on to the sky. We can travel to anywhere on Earth, and show the sky as it appears from Australia or Antarctica; we can travel backwards and forwards in time, and show the sky as it was in the time of Christ, or as it will be in AD 12,000, when the north celestial pole will be close to the brilliant Vega rather than our modest Polaris.

The first modern-type planetarium was set up at Jena, in Germany, in 1923, mainly by the famous optical worker Walther Bauersfeld. Subsequently they proliferated: the London Planetarium was opened in 1961 (I declined an invitation to be its first Director), and in 1966 came the Armagh Planetarium, which I did take on and ran for 3 years. But more were needed, particularly as the London Planetarium at Madame Tussaud's is a commercial venture rather than being purely scientific. Chichester, in Sussex, seemed to be a good venue.

The South Downs Astronomical Society was very active, and one of its leading lights, Dr John Mason, proposed that it should establish a planetarium. He enlisted helpers, mainly John Green, Peter Fray and Roger Prout, and we began to investigate the possibility.

▼ The projector of the South Downs Planetarium, which we acquired from Armagh.

This was where I could play a role. When I became Director at Armagh, I had chosen an excellent Japanese projector, and had even travelled to Tokyo to design the star plates. It worked well, but in the 1990s the decision was made to replace it with a Digistar, which is more 'modern' and will do tricks, but in my view is not nearly so good – the star images are much less crisp, and there is no colour. However, the change meant that Armagh had no further use for its Japanese projector, and we were able to obtain it. The Armagh authorities could not have been more considerate or more helpful; they sold us the projector at a very modest price. But for them, the South Downs Planetarium would have been stillborn.

When we had the projector, there was little money left, and we had to do some frantic

▲ **The Astronomer Royal,** Sir Martin Rees, *with me at the opening of the Planetarium.*

fund-raising. Local industry was helpful, and the local council helped. We had pinned our faith in the Lottery; this was an educational project, with special emphasis upon the young people who would be the country's mainstays in the future. Surely the Lottery would help us? Not so. They sent a blank refusal. We were some years away from the next General Election, so I doubted whether our Conservative MP would be interested; he wasn't. Neither were the Labour and LibDem candidates, who did not even answer. Of the political parties, only UKIP helped, both by giving us full support and paying a subscription.

Otherwise we had to 'go it alone'. We did all the usual things; for example we gave public lectures, held coffee mornings, and so on – but these were clearly limited. I did break my usual rule and did one harmless TV advertisement, which was actually very well paid – naturally, the whole fee went to the Planetarium.

One decision had to be made, and in my view we made the wrong choice. We were offered a site and a building just at the edge of the grounds of the Chichester Boys School, and this was accepted. Agreed, we were able to set up a really good Planetarium, which would otherwise have taken us years, but the snag was that it made us look like a purely schools Planetarium, which we are not. If we had an entrance off the main road all would be well, but we haven't; to get to the Planetarium you have to drive or walk through part of the school grounds, and this gives a wrong impression – though, I hasten to add, school audiences make up a very important part of our programme. Too late now, but I wish we had been entirely separate.

The projector had not been used for years, and needed drastic overhauling. Luckily John Mason was an expert at this sort of thing; without him, we would have been in very serious trouble, to put it mildly. With an immense amount of work he restored the projector. We bought a dome, and erected it in the building. We were lucky in being able to buy aircraft tip-back seats (again from Ireland). We were given a computer, and some books to set up a library. The Planetarium itself began to take shape; again local industry was amazingly generous and helpful. At last came the great day. Who to officiate but the Astronomer Royal, Sir Martin Rees, who had been so supportive throughout! He came; it was a day not to be forgotten and we were in business.

John Mason has done virtually all the lecturing, and he could not be bettered. John Green has spent most of his time at the Planetarium, and without him we could not have managed (it was fitting that in 2004 he was given the MBE; John Mason ought to have had one too). Roger Prout and Peter Fray were also very active, and Roger, who is a 'legal eagle', coped with all the finance.

Sine we opened, many thousands of people have been to the Planetarium, and meetings too have been held there; I know it is doing a really good job. We need to employ staff, but as yet we haven't enough money; the Planetarium maintains itself, and makes a modest profit, but all this is ploughed back into making improvements. We will press on, and hope that in the future we will be able to extend our scope.

Well, the Planetarium is a success, and we have the satisfaction of knowing that all the years of planning and hard work have been worthwhile. If you are within range, do come and see us. You will find us at the South Downs Planetarium, Kingsham Farm, Kingsham Road, Chichester P019 8RP; we are on the phone – Chichester 774400. So if we do welcome you here, I hope that you will enjoy yourselves under our artificial but immensely realistic sky.

11 • TARGET – EARTH!

Anyone opening a daily paper towards the end of July 2002 might have been forgiven for feeling slightly alarmed. Here are some of the headlines:

THE END IS NIGH (*Daily Mail*, 25 July)
INVADER: MASSIVE ASTEROID TO HIT EARTH (*The Sun*, 26 July)
MILE-WIDE ASTEROID POSES GREATEST THREAT YET (*The Times*, 26 July)
NEVER MIND IRAQ: DESTROY THE MONSTER FROM OUTER SPACE (*Daily Express*, 28 July)
THE WORLD ENDS ON FEB 1ST 2019 (POSSIBLY) (*The Daily Telegraph*, 25 July)

The cause of the furore? An asteroid, 2002 NT7, discovered on 9 July by astronomers at America's Linear Observatory in New Mexico. Its diameter is about one mile, and it moves round the Sun in an orbit which takes it from around the distance of Mars to just inside the Earth's orbit; the period is 837 days. The orbit is tilted with respect to ours, but the preliminary calculations indicated that there was a real risk that the asteroid would hit Earth on 1 February 2019. The effects of such a strike would be devastating, and the Press conjured impressions similar to those in films of the *Armageddon* type.

In fact, it was soon found that NT7 would miss us by a wide margin, and would not even approach us more closely than the orbit of the Moon. But it is far from unique; there are many hundreds of NEAs or Near-Earth Asteroids, to say nothing of comets. We are not immune, and it would be foolish to pretend otherwise. Indeed, on 14 June 2002 a smaller asteroid, 2002MN, brushed past at a mere 120,000 km (75,000 miles), and in 1994 another small asteroid came even closer than that.

There is a widely supported theory that a vast impact, 65 million years ago, caused global devastation and changed the entire climate – with disastrous results for the dinosaurs, which had ruled the world for so

▲ *2002 LV, a Potentially Hazardous Asteroid,* is seen to move against the background stars.

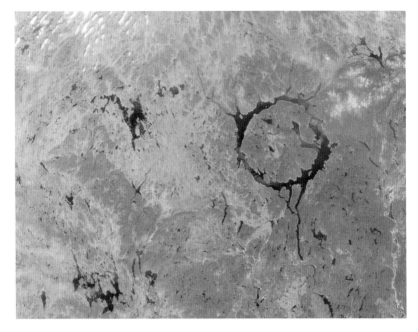

▲ **Lake Manicouagan,** in Quebec, Canada, is thought to have been formed 212 million years ago by the impact of a 5-km (3-mile) wide asteroid.

long but were quite unable to adapt to the new situation. The crater produced at that moment is thought to be at Chicxulub, off the coast of Mexico. Nothing as catastrophic as that has happened since, at least as far as we know, but in 1908 a missile struck the Tunguska region of Siberia and blew pine trees flat over an area of 2000 square kilometres (700 square miles), and in 1947 another sizeable object came down in the Vladivostok area, breaking up before impact and producing a swarm of craters. If either of these bodies had hit a city the death toll would have been colossal.

Impact craters are not uncommon; the most celebrated is Meteor Crater in Arizona, described by the Swedish scientist Svante Arrhenius as 'the most interesting place on Earth'. It was formed 50,000 years ago, long before the area was inhabited. Another well-formed impact crater is Wolf Creek, in Australia, and it has recently been established that there is an impact crater – or rather, indications of it – on the bed of the North Sea. The Moon, of course, is covered with impact craters, which look fresh because of the lack of erosion; there is to all intents and purposes no atmosphere around the Moon. And in July 1994 a comet, Shoemaker–Levy 9, plunged into Jupiter and caused effects which persisted for many months. Mind you,

Jupiter is a very large target compared with our tiny Earth.

In 1998, at a meeting held at Turin in Italy, the International Astronomical Union produced the 'Torino Scale' with respect to possible asteroid or comet impacts. It works on a scale of 0 to 10. 0 means don't worry; 1, keep watch; 2 to 4, cause for some concern; 5 to 7, cause for grave concern; 8 to 10, prepare for something like Armageddon. It is worth giving this scale in rather more detail:

0. *Virtually no chance of collision.*
1. *Chance of collision very unlikely.*
2. *Fairly close encounter, but little likelihood of collision.*
3. *Close encounter; slight chance of localized damage.*
4. *Close encounter, with a real chance of regional damage.*
5. *Significant threat of regional damage.*
6. *Significant threat of global damage.*
7. *Very real threat of global damage.*
8. *Threat of localized destruction.*
9. *Probable regional devastation.*
10. *Global catastrophe.*

A Torino 10 event is unlikely to happen more than once in about 100,000 years, but of course one never knows! There is also the Palermo Scale, which differs from the Torino and is used by specialists to evaluate the concern warranted for a future potential impact possibility.

All NEAs are small, but travel at high velocities, and they do constitute a direct threat – now being recognized by some (not all) politicians. Teams of astronomers in the United States and in Japan are doing their best to locate threatening wanderers, and it is to be hoped that before long Britain will join in. But if we did locate a likely impactor, is there anything that we could do about it?

For a body much more than a mile across, the frank answer is 'No'. But we might be able to cope with a smaller object, provided that we had sufficient warning. An

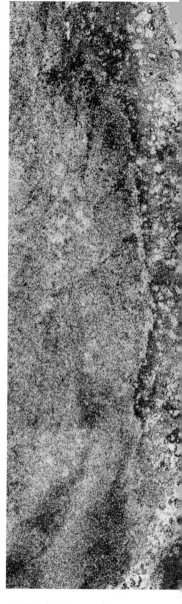

▲ *Chicxulub impact crater,* Mexico, imaged by radar. The southwest portion is seen here.

immediate problem is that in many cases NEAs escape detection until they are almost upon us, or have already passed their point of closest approach.

The instinct is to say 'Destroy the asteroid with a nuclear bomb'. But this would not be a good idea, even if it were practicable. The result would be to turn the asteroid into lethal cosmic shrapnel. So the only real solution is to divert it, and make it miss the Earth altogether. A tiny induced change in the asteroid's orbital velocity would do the trick – if we had enough warning and if we had sufficient power at our disposal.

It is said that when someone asked Napoleon to do him a favour, the reply was 'I will give you anything except Time'. This is certainly

◀ *My asteroid,* No. 2602 Moore, imaged by John Fletcher. I lie in the outer part of the main belt, so there is no fear that I will collide with the Earth!

▼ *Meteor Crater, Arizona,* is about 1500 m (5000 ft) wide. It was created by the impact of a 50-m (160-ft) asteroid.

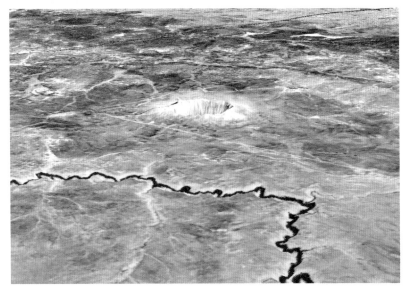

▶ **Potentially Hazardous Asteroid 1998 WT24.** *This 10-minute exposure was guided on the asteroid, which appears as a white dot among the star trails.*

true of asteroid diversion. We would need many months at least, which means that the asteroid (or comet) would have to be identified well ahead of the estimated impact date.

One idea, seriously investigated by NASA, involved fixing a nuclear engine to the asteroid. The engine would be flown to the asteroid on a conventional rocket; once firmly on the surface, it would fire and literally push the asteroid off course. This might work with sufficient time, but only for a small body; a mile wide asteroid such as NT7 would be too much of a problem. Fixing a 'solar sail' on to the asteroid's surface, and relying upon light-pressure from the Sun, would be too weak. All in all, the most promising method seems to be to cause a nuclear explosion some way from the asteroid. This would unleash a shock wave of radiation which would vaporize material on the asteroid's surface, and cause a slight but vitally important change in its orbit.

Obviously we need to know how solid asteroids are. Some, such as Eros, are certainly monoliths; others, such as Mathilde, are fragile 'rubble piles'. The asteroids make up a heterogeneous family.

Will we be hit? Probably. Impact scars are not uncommon – only this month (August 2002) an undersea crater, Silverpit, was located on the bed of the North Sea, nearly 150 km (90 miles) from Hull. It seems to be around 65,000,000 years old, contemporary with the Chicxulub impactor which allegedly wiped out the dinosaurs, and will have been formed by an asteroid about 120 metres (400 feet) across. The crater itself has a diameter of over 3 km (2 miles), and is surrounded by a series of concentric rings extending out for 8 km (5 miles) in each direction, a tall central peak is buried inside the crater. The general pattern looks very similar to some of those on Jupiter's icy satellite Europa.

At least we are now on the alert. If we do have advance warning, then we may well be able to avert disaster, and certainly we are better prepared than the dinosaurs were.

— 12 • NEW LIFE AT JODRELL BANK —

When I first saw Jodrell Bank, it really was a grassy bank – no more. Then it was transformed into the most famous radio astronomy observatory in the world, and the whole scene is dominated by the huge dish of the 76-metre (250-ft) Lovell Telescope. Jodrell Bank is still in the forefront of research – and yet not so long ago there was a serious threat that it might be closed, or at least the Lovell Telescope would be taken out of service. Accountants, those parasites of the modern world, were looking for financial cuts. You can't tax the Lovell Telescope, so clearly it was vulnerable!

You can well imagine the outrage at this suggestion. Led by the Director, Professor Andrew Lyne, the scientists fought back, and on this occasion they won. Funds were given – and, more, there was

▼ **The Lovell Telescope,** at the Jodrell Bank Observatory, Cheshire, UK. *It came into use in 1957 – just in time to track Russia's Sputnik 1.*

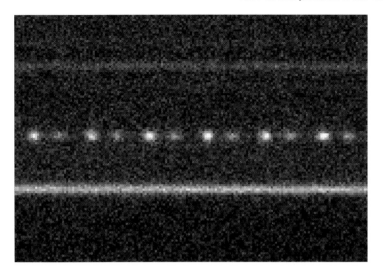

▲ *Time sequence of a pulsar.* In this image the continuous lines at top and bottom are produced by stars of constant brightness. The dots are produced by the pulses of the Crab Nebula's pulsar.

money (around £2 million) for modernizing the Lovell Telescope, which was in dire need of overhaul. So the future of Britain's great observatory seems secure – at least for the moment.

On its own the Lovell is immensely sensitive, but even more so when working with other radio telescopes. It is part of MERLIN, which stands for Multi Element Radio Link Inteferometer Network, and has absolutely nothing to do with an Arthurian wizard! There are seven radio telescopes in MERLIN, two at Jodrell (the Lovell, and the Mark II 26 metres/85 ft) and five in various other sites; the furthest out is at Cambridge. These are now being connected with fibres, which is a vast improvement on the older methods.

I remember well the 30th anniversary of the Lovell Telescope, in 1987. There was a major celebration here, and the figures 57–87 were painted on the dish; one of my tasks was to help in decoying Bernard Lovell away until the great moment of 'unveiling'. I am glad to say that we succeeded!

Pulsars have always been a special interest of the Director of Jodrell Bank, Andrew Lyne. A pulsar is a stellar wreck – all that is left of a star which has exploded as a supernova, and thrown most of its material away into space; the remnant is very small, incredibly dense, and spinning round rapidly, emitting radio pulses. The most famous pulsar in the Crab Nebula (the remnant of the supernova of 1054) pulses 30 times a second. There are also 'millisecond pulsars', which have been 'spun up' and now spin thousands of times per second. There are at least 20 in the globular cluster 47 Tucanae, spinning at various speeds. Of course, no conventional noise can come from space – and 47 Tucanae is 15,000 light-years away; it contains around a million stars. The noise is created in the radio receiver, but it is fascinating to hear what notes would be recorded – truly, 'the music of the spheres'.

One discovery made by a Jodrell Bank astronomer, Dennis Walsh (though he was not actually at Jodrell at the time), was that of gravitational lenses. Einstein pointed out that light rays can be bent if passing close to massive bodies, and it was then realized that concentrations of matter, such as stars and galaxies, could act as distorting – 'gravitational' lenses, focusing the light from objects behind them. This can be shown by taking the stem of a wine-glass (I'm afraid you will have to break it away from the glass bowl!) and looking through it on to marks on a piece of paper.

In 1979 Walsh and his team had begun a survey of radio sources, using the Lovell Telescope, and linking them with observations at optical wavelengths made in Arizona. One source seemed to consist of two quasars, very close together – just sufficiently far apart to be recognized through the atmospheric blurring. The spectra were showing Walsh that he was dealing with two images of the same quasar. Subsequently, a massive galaxy was found along the line of sight from the Earth to the quasar images. This was the first gravitational lens; many are now known. One splendid example is of the radio source 1938+666, where the radio map shows lovely arcs and the Hubble Space Telescope shows the background galaxy distorted into a complete 'Einstein Ring', with the lensing galaxy in the middle

▶ *Einstein ring,* from the Hubble Space Telescope. The distant galaxy 1938+666 has been distorted into a ring by the intervening galaxy. This 'lensing galaxy' appears as a bright spot in the middle of the ring.

looking like a bulls-eye. Here, the alignment is exact; if not there will be luminous arcs rather that a compact ring.

Gravitational lenses are important because they make it possible to take a census of the amount of matter in the universe, both light and dark. We can also use the distortions to give information about how matter is distributed, and we can use the numbers of lenses to increase our knowledge of the overall geometry of the universe. We see bright matter, we infer the presence of dark matter, but it seems now that the universe is dominated by 'dark energy', involving the Cosmological Constant – introduced by Einstein, but later rejected by him. This dark energy has anti-gravity-like properties, and the universe is accelerating even faster over periods of time rather than being slowed down by gravity.

But what is the nature of dark energy? To be honest – nobody has the faintest idea.

Radio astronomers are always calling for 'more sensitivity' just as George Ellery Hale once called for 'more light'. We need larger radio telescopes – larger even than the 300-metre (1000-ft) dish at Arecibo in Puerto Rico, built in a natural hollow in the ground. One exciting new venture, now in the planning stage, is the Square Kilometre Array, or SKA.

The size is regulated by the need to study hydrogen gas in remote galaxies. Hydrogen, the most abundant element in the universe, emits radio radiation at a wavelength of 21 centimetres. This makes it possible to study the movements of gas within galaxies under the action of gravity, and Dutch astronomers found that galaxies contain dark matter as well as bright material. Unfortunately the 21-centimetre radiation is weak, limiting us to studies of relatively nearby galaxies. To study galaxies in the far reaches of the universe, formed in the first thousand million years after the Big Bang, we need a telescope with a collecting area of at least 1 square kilometre – 200 times that of the Lovell Telescope.

A dish of this size cannot be built, and moreover the SKA is too big a project to be the responsibility of any one nation; it must be international. It will be made up of numbers of small radio elements spread out over hundreds of kilometres; these telescopes will be connected by optical fibres. The precursor is the Allen Telescope Array (ATA) now being developed in California; it is made up of over 300 radio telescopes, each 6 metres in diameter.

Another possible method is to make the SKA not out of 'dishes' but out of dipoles, on which our rooftop TV aerials are based. These can be connected electronically, and the results fed into a giant computer. Such a 'phased array' of telescopes has the advantage of being able to look in many directions simultaneously. At present radio astronomers are blissfully unaware about what is happening in most areas of the sky, and a multi-beam telescope would allow us to keep a constant watch for temporary, unexpected phenomena, such as the

▲ **The Arecibo Telescope,** in the Guarionex Mountains of Puerto Rico. It is built into a natural hollow in the ground.

mysterious GRBs or gamma-ray bursters – titanic explosions whose nature is still unclear.

One problem is that the expansion of the universe stretches the 21-cm wavelength to over a metre in length, and here there is a very strong interference from commercial transmissions. The SKA will therefore have to be built in a very remote area, away from populated regions. The best of all possible sites would be the far side of the Moon, which would be completely radio-quiet.

This will no doubt be possible later in the 21st century. Meanwhile, we must manage with what we have. Jodrell Bank has led the way in the past; now that its future is secure, it will continue to take the lead in the years to come.

13 · DOBSONIANS

Telescope mounts are of many kinds. Some are simple, others very elaborate. Today the tendency among amateurs is to go in for complicated mountings, so that the telescope can be accurately aimed and driven – essential if devices such as CCDs are to be used. But there are many ways and means, and the simplest of all mounts is unquestionably the Dobsonian.

The mounting is the brainchild of John Dobson – an amateur astronomer, who studied chemistry and spent a period as a monk. It really is basic. The mirror is mounted at the bottom of the tube, which may be made of almost anything and which can be any shape (square Dobsonians are common!). The tube is rigid, and the telescope can be swung up and down inside it. The tube itself is fixed to a turntable, giving full rotation. Everything else is 'normal' enough. Clearly the design is best suited to a Newtonian, though other optical systems cannot be ruled out.

Of course the vital part of the telescope is the main mirror. It can be bought, but skilful amateurs can make excellent mirrors of large size. John Dobson himself used to grind mirrors from discarded ship's portholes and the bottoms of glass jugs!

The principal advantage of the Dobsonian is its portability. Would you like, say, a 24-inch (600-mm) reflector? Conventionally mounted, this involves massive components and almost certainly a permanent observatory, but a 24-inch Dobsonian can fit into a car. I even know of a couple of 36-inch (900-mm) Dobsonians, owned by

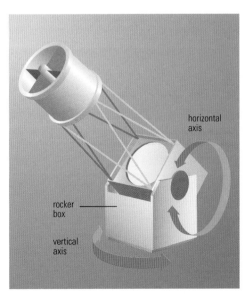

horizontal axis

rocker box

vertical axis

▶ *The Dobsonian design, which has enabled many amateurs to build their own large-aperture instruments. The great advantage is their portability.*

friends who live in a light-polluted area and have to escape into the darkness of the countryside.

The second main advantage is the ease of operation. There is so little to go wrong; also, the eyepiece is always comfortably available.

The chief disadvantage is that an ordinary Dobsonian cannot easily be fitted with mechanical slow motions. You simply have to push the telescope around, and this means that everything has to be carefully made, so that the movements are smooth. Backlash and similar problems must be avoided at all costs. Even so, using high magnifications poses real problems, and keeping an object such as a planet in a small field of view is frustrating. It is fair to say that a Dobsonian is above all a light-bucket. For wide fields and low powers it is superb; the sight of, say, a globular cluster when seen through a large Dobsonian is breathtaking.

14 • THE ACCELERATING UNIVERSE

Of all the questions which have been asked over the ages, one of the commonest is: 'How old is the universe?' It was suggested that the universe might only have been born a few thousand years ago. We know better now; we are confident that the true age of the universe, as we know it, is around 13.7 thousand million years. We also know that the universe is expanding; all the groups of galaxies are racing away from all the other groups – and the further away they are, the faster they are receding.

Our main evidence comes from the Doppler effect. All the lines in the spectra of galaxies (apart from those in our Local Group) show redshifts. It has been suggested that these shifts are due to causes other than the Doppler effect, but most astronomers (not all) now believe that other explanations do not stand up to careful analysis. Fred Hoyle and others championed the 'steady-state' theory, according to which the universe has always existed and will exist for ever, but during the 1960s came several new pieces of evidence. According to the steady-state theory, the far parts of the universe would look the same as the closer parts – but they do not. Then there was the discovery of the cosmic microwave background, the afterglow of the Big Bang. (Ironically, this term was introduced contemptuously by Hoyle himself!)

But how did the Big Bang happen? We have to admit that we do not really know. We can go back to a tiny fraction of a second after the Big Bang, but conditions before that – with the immense densities and temperatures – are beyond our present-day physics, and remain cloaked in mystery.

Modern telescopes can show us objects as they were when the universe was no more than 10 per cent of its present age, and in

▶ **Brown dwarf 2M1207 and its planet.** *Ten years after the discovery of the first extra-solar planet (and two years after this* Sky at Night *episode was broadcast), astronomers at the Very Large Telescope managed to image a giant planet, orbiting a young brown dwarf star. It is near the constellation of Hydra.*

778 mas
55 AU at 70 pc

▲ *Supernova 1994D* *(lower left), in the galaxy NGC 4526, imaged by the Hubble Space Telescope. Astronomers can use such events to measure distances to far-away galaxies, which in turn enables them to work out the age of the universe.*

this respect astronomers are luckier than geologists; they can actually look back in time, whereas geologists can only see things as they are now. But each advance raises a whole host of new problems. For example, if the rule of 'the further, the faster' holds good, we will come to a point when a galaxy is receding at the full speed of light, and will be unobservable; this may limit the size of the accessible universe, but how do we know that this is the full story? A sailor at sea may have his view limited to the horizon around him, but he will have no means of knowing how much horizon lies beyond. And the universe is probably much larger than the portion available to us.

There is also the question of the rate of expansion. We know that there is a great deal of 'dark matter' which we cannot see, and whose nature is uncertain; there is also 'dark energy', which is a total mystery, and is latent in space. Yet it does not seem that there is enough matter, dark matter and dark energy to slow down the rate of expansion, and indeed it seems that the universe is accelerating. Studies of remote supernovae, in particular, indicate that the rate of expansion was slower in the past than it is now. Einstein, in 1917, believed the universe to be static, and to halt any expansion he introduced a force called the cosmical repulsion, acting against gravity. When Hubble proved that the universe really is expanding, it became clear that Einstein had missed an opportunity – but we now find that he may have been right after all.

It seems that the universe is made up of 5 per cent atoms, 25 per cent dark matter and 70 per cent dark energy. If the acceleration is real – and several lines of investigation support it – then the universe will become ever colder and emptier. The Earth cannot survive for more than around 6 thousand million years before the Sun destroys it, but the universe itself has a much longer future – perhaps even an indefinite one.

At least we are starting to understand how stars form and evolve, and we are trying to find out the story of the universe itself – how it developed from the initial Big Bang. There are other fascinating problems to be faced, too. In 1995 we discovered the first planet of another star (51 Pegasi); by now over 100 are known, and within 10

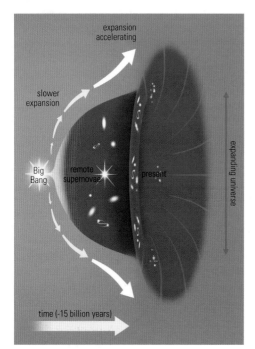

◄ **Accelerating universe.** *Most astronomers now believe that the rate of expansion of the universe is increasing.*

▼ *Type Ia supernovae* are believed to occur in binary star systems. Hydrogen leaks on to the white dwarf component of the system. When the dwarf's mass increases above a critical limit it explodes as a supernova.

years we may have identified many more. There may well be other Earths, with life; why should we be unique? We must search, using modern-type equipment and space research methods; the discovery of life elsewhere would indeed be an epochal event, and it could happen. Instruments today are of amazing power – strange to remember that less than a century ago the largest telescope in the world was the 72-inch (1.8-metre) Rosse reflector at Birr Castle in Ireland. With the detectors we use now, a modest 6-inch reflector when used photographically is as effective as a 60-inch used to be.

There will also be the New Generation Space Telescope – now named the Webb Telescope, which may take us back to the time when the first stars 'lit up' following the dark, starless period immediately after the Big Bang. At least we may be sure of one thing: the coming years are going to be very exciting!

15 • OUR TURBULENT SUN

Our programme is called *The Sky at Night*, but sometimes we have to discuss the Sky by Day – because our Sun, which seems so splendid to us, is nothing more than a very ordinary star. It is, however, a mere 150,000,000 km (93,000,000 miles) away, and the nearest star beyond the Sun lies at 39 million million km (24 million million miles).

The Sun is a globe of intensely hot gas, mainly hydrogen and helium, which pours out light, heat and other forms of energy into space. It would contain more than a million globes the volume of Earth; its overall diameter is about 1,400,000 (865,000 miles) – a hundred times that of the Earth. But if you could place the Sun in one pan of a pair of scales, you would need no more than 330,000 Earths to balance it. It is less dense than the Earth, and on average only slightly denser than water. Its visible surface, known as the photosphere, is at

▼ **The Sun,** imaged by SOHO (the Solar and Heliospheric Observatory). Sunspots are clearly visible.

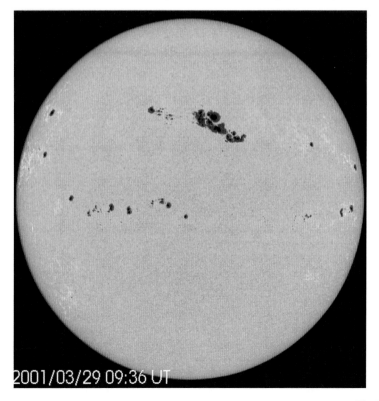

2001/03/29 09:36 UT

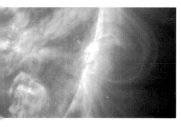

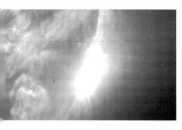

▲ *A solar flare,* as captured in a series of images taken by the SOHO spacecraft.

a temperature of about 5500°C, but near the core the temperature rockets to something like 15 million degrees.

The Sun shines because of nuclear reactions in the central core. At such a high temperature, atomic nuclei are moving very quickly and colliding with each other; in these collisions nuclei of the lightest element, hydrogen, are being welded together to form nuclei of helium. Each complete reaction converts four hydrogen nuclei into one helium nucleus, but the resulting helium nucleus weighs very slightly less than the original four protons (hydrogen nuclei), and this difference is released in the form of energy. As Einstein showed, a small amount of mass converts to a huge amount of energy. For example, if you could turn just one kilogram of matter completely into energy, you could produce enough energy to keep a 1-kilowatt electric heater running for about three million years. To maintain this huge output of energy, the Sun has to destroy four million tons of matter every second.

The energy generated in the core flows out in the form of photons – tiny particles of energy which battle their way outwards through what is called the radiative zone, suffering thousands of millions of collisions with particles of matter en route. About 70 per cent of the way from the centre to the surface, the outflowing energy enters the convective zone, in which energy is carried to the solar surface by the process of convection. Huge currents of hot gas rise to the surface; they radiate energy into space, and then sink down to be heated again.

Almost all the Sun's visible light is released into space through the thin layer of gas called the photosphere. Below the photosphere the Sun's material is opaque to light, but above the photosphere it is practically transparent. In fact, the photosphere is the visible surface of the Sun.

Two principal layers of atmosphere lie above the photosphere: the chromosphere and the corona. The chromosphere is only a few thousand kilometres deep, whereas the corona is very extensive, stretching out to many times the diameter of the visible disk. Although both layers emit light, they are so drowned by the brilliance of the photosphere – or, for that matter, a clear blue sky – that they cannot be seen with the naked eye except during a total solar eclipse, where the Moon passes in front of the Sun and temporarily blots it out. This happened on 4 December 2002, as seen from parts of Africa and Australia. The chromosphere and corona were beautifully displayed. With instruments based upon the principle of the spectroscope the chromosphere can be studied at any time, and the corona can be seen by means of coronagraphs – instruments that incorporate a special disk to blot out the photosphere. And of course, we can now use instruments carried in spacecraft, such as SOHO, the Solar and Heliospheric Observatory.

The corona is at an amazingly high temperature. If we look at how the temperature of the solar atmosphere changes with height above the photosphere, we find that it first falls to around 4000°C and then rises very sharply across the 'transition region', rising to between 1 and 5 million °C in the corona itself. Just why this happens is still unclear.

Though the coronal temperature is so high, there is relatively little 'heat'. Temperature depends upon the rate at which the particles (protons and electrons) are moving around. In the corona the speeds are very great, but the density is very low – a million million times less than that of our air. This is why the corona sends out so little heat.

Sunspots are regions in the photosphere which are cooler than their surroundings; any telescope is capable of showing them. (Never look direct; the safest method is to use the telescope to project the Sun's image on to a screen.) The dark central part of a spot, the umbra, is from 1500 to 2000 degrees cooler than the adjacent photosphere, and so less light is emitted, though if a spot could be shining 'on its own' it would still be brilliant. A typical spot has a dark umbra embedded in a lighter region known as the penumbra; often there are many umbrae in one mass of penumbra.

Spots generally appear in groups – regions of concentrated magnetic fields where the magnetic forces are thousands of times stronger than the Earth's field. In a typical spot-pair, one spot will behave like a north magnetic pole and the other like a south magnetic pole; it is almost as though a bar magnet lies below the surface between the two spots. The orientations and strengths of the magnetic fields are described by imaginary lines, called field lines, which link one pole with the other.

In the case of a spot-group, magnetic field lines have been crowded together, concentrating the magnetic field, until rising in a loop through the surface, one spot – with positive magnetic polarity – forms where the field lines emerge from the surface, and the other – with negative polarity – forms where the lines plunge back below the surface.

By using spectral analysis, solar physicists can produce images known as magnetograms, which show where the surface magnetic fields are concentrated. Regions of concentrated magnetic fields are shown as bright patches (positive polarity) or dark patches (negative). When we compare a magnetogram with a white-light view, the positions of the sunspot groups clearly correspond to the underlying regions of concentrated magnetic field.

Cutting out all the light except that due to hydrogen, we can study the chromosphere directly; we can see flame-like projections of gas round the edge of the Sun's disk (prominences) and long, dark straggly features (filaments). Prominences and filaments are clouds of relatively dense gas suspended above the solar surface. When such a cloud lies between us and the photosphere below, it appears dark because it is absorbing the light from the photosphere; when it projects beyond the limb it is bright, because we can then see the light emitted by the hot hydrogen against the blackness of space. Prominences and filaments are suspended in the Sun's atmosphere by magnetic forces, and are often to be found along the boundary between regions of opposite magnetic polarity. The weight of the gas in the prominence – which can be a hundred times denser than the surrounding corona – is supported by magnetic field lines.

Eruptive prominences can surge outwards at speeds of hundreds of kilometres per second, though quiescent prominences can hang almost unchanged for many days. All are of vast size compared with the Earth.

A flare is the sudden and dramatic release of pent-up energy, stored in the twisted magnetic fields associated with complex spot-groups and magnetically active regions. Given about twenty minutes or so, a major flare can release up to 10^{25} joules of energy, equivalent to the energy released by the simultaneous explosion of several thousand million one-megaton hydrogen bombs. A flare releases radiation at all wavelengths, from gamma rays to radio waves, but mainly as X-rays. Subatomic particles such as protons and electrons are accelerated to nearly half the velocity of light, and masses of high-temperature plasma (a mixture of protons and electrons) are sent hurtling outwards, sending shockwaves through the corona and across the face of the Sun.

It seems that flares are triggered by a process known as magnetic reconnection. Magnetic field lines behave rather like elastic strings; when they are stretched and twisted they contain a great deal of energy which can be suddenly released. If an elastic band is stretched and then suddenly released, it snaps back and is quite capable of propelling a paper pellet across the room! In a similar way, if oppositely directed magnetic field lines are pushed together by the movement of hot plasma in the chromosphere or corona, oppositely directed lines can come into contact, releasing vast amounts of energy as they do so. Plasma at the point of reconnection is heated to tens of millions of degrees, hot enough to give off X-rays, charged particles are accelerated along the field lines, and blobs of plasma are catapulted out through the corona.

Magnetic reconnection also seems to play a major role in heating the corona. Large numbers of small-scale reconnection events are taking place all the time, each one heating plasma locally and contributing to the heating of the corona as a whole.

Magnetic forces drive all aspects of solar activity. They determine the shape and structure of the corona, and play a major role in perhaps the most dramatic of all solar events: coronal mass ejections (CMEs), in which vast blobs and shells of plasma, with intertwined magnetic fields, are catapulted out through the corona into space. Each CME contains around a thousand million tons of plasma, and

▼ *A huge prominence,* imaged by SOHO, arches above the Sun. With very large prominences, material sometimes escapes from the Sun completely.

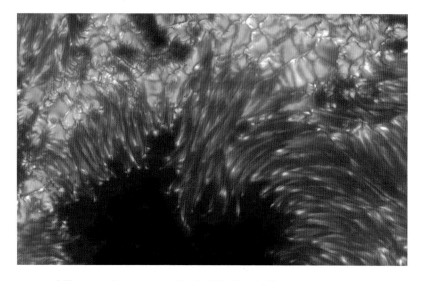

▲ The edge of a sunspot, in a spectacular image captured by the Swedish Solar Telescope. The dark umbra is surrounded by the brighter penumbra.

they hurtle outward at speeds ranging from 200 to over 2400 kilometres per second (125 to 1500 miles per second). The underlying field lines control the way in which the plasma flows; the shapes of CMEs can be very convoluted.

Plasma is escaping from the corona all the time, in what is called the solar wind. Where closed loops of magnetic field in the corona are strong enough, they trap the million degree plasma and prevent it from escaping. However, in other parts of the corona the field lines do not form closed loops – that is to say, they are open – and hot plasma can readily escape. These 'leaky' areas of the corona are known as coronal holes.

The particles making up the solar wind are mainly protons and electrons. They flow out through the Solar System, blowing like a wind past the planets at around 400 to 520 kilometres per second (250 to 320 miles per second). This stream of electrically charged particles, carrying magnetic field lines with it, interacts with the magnetic fields of the planets, and plays a major role in dragging plasma away from the heads of comets to form their ion tails. The Earth has a magnetic field, which in some respects acts as though there were a bar magnet embedded in the molten core. The fast-flowing solar wind squeezes the Earth's magnetic field inwards on the Sun-facing side and draws it out into a long tail (the magneto-tail) on the opposite side. The region in which the Earth's magnetic field is dominant – that is to say, stronger than the interplanetary field – is known as the magnetosphere. It acts as a shield, causing the solar wind to flow round the outside of the magnetosphere rather that ploughing straight into our atmosphere.

Charged particles from the Sun are also responsible for aurorae. A major outburst such as a flare or a CME sends out a burst of energetic particles into space – a major gust in the solar wind. When these particles reach the Earth, they squeeze and distort the magnetosphere, causing electrical currents to flow. This produces magnetic storms, which deflect compass needles, cause power surges in electric cables and telephone lines, and cause charged particles to plunge down magnetic field lines in the vicinity of the Earth's north and south magnetic poles. The particles interact with atoms and molecules in the upper atmosphere, making them emit light and producing aurorae. Aurorae tend to occur in oval bands, centred on the magnetic poles, and so are best seen from high latitudes, though after a major solar eruption the auroral bands expand to produce displays visible much closer to the equator. It is on record that one aurora has even been seen from Singapore!

The Sun has a cycle of activity of around 11 years; the numbers of spots and groups varies in a fairly regular way. Around solar maximum the photosphere is heavily spotted, while near solar minimum there may be days or weeks with no spots at all. Near spot maximum, magnetograms show how the magnetic fields are concentrated into two bands of bipolar magnetic regions, one on each side of the Sun's equator; at minimum there is no obvious concentration of magnetic fields.

The last maximum fell between 2000 and the beginning of 2002. It was unusual inasmuch as there were two distinct peaks; the first major peak occurred around March 2000 and the second around December 2001. The next minimum is expected around 2006.

We owe everything to the Sun – but it will not last forever. Though the core contains vast reserves of hydrogen fuel, enough to maintain it for another 5000 million years, the supply will eventually run out. When the core has used up its available fuel, it will shrink; as it does so its temperature will rise, and hydrogen will start to 'burn' in a shell surrounding the original core. In consequence the Sun will actually become brighter, and will swell up to become a huge distorted red giant, 40 to 50 times larger than it is at the moment. The temperature of the Earth's surface will soar; the oceans will evaporate, surface rocks will melt and the atmosphere will be driven off into space, so that life here will no longer be possible. A few hundred million years later the Sun's nuclear furnace will switch off, the outer layers will be puffed out into space, and what is left of the Sun will shrink down to become a tiny white dwarf star no larger than the Earth. Finally, after thousand of millions of years, the Sun will fade into oblivion. But at least nothing of the sort is likely to happen yet awhile; we have plenty of time to prepare!

—— 16 • ASTRONOMY AND ART ——

Nobody can have any doubt about the beauty of celestial objects. Nothing on the Earth can match the glory of Saturn's rings, the coloured nebulae or the graceful spiral galaxies, so far away that their light takes millions of years to reach us. Pure 'space art' is often thought to be a fairly modern development, but in fact it goes back a long way.

One of the pioneers was James Nasmyth, author – or co-author – of the first detailed book about the Moon to be published in English; as an aside, he was also the inventor of the steam-hammer! With James Carpenter he worked out a theory of the origin of the lunar craters, which has been remembered as the 'volcanic fountain' theory and which seemed at the time (1871) to be convincing, though we now know that it was completely wrong. Nasmyth pictured a volcano in violent eruption, sending out material which formed the crater wall. He produced skilful drawings showing how, in his theory, a crater was gradually built up, and he also published remarkably accurate pictures of craters modelled in plaster.

However, it is probably true to say that the real 'father of space art' was an American, Chesley Bonestell, whose work was very influential from the middle of the 20th century. His paintings of, for example, the landscapes of Saturn's satellites are vivid, and were based upon the facts as they were then known – and on the whole they were surprisingly accurate. Bonestell had the advantage of working with some

▼ **Saturn,** drawn by Paul Doherty with my 15-inch (380-mm) reflector at home in Selsey.

▲ *Future Martian base?* In February 2003 we asked viewers to submit their ideas for a base on Mars. Reproduced above is Frank Norrington's entry.

of the space pioneers, notably Willy Ley, who came to America when the Nazis began to take control of science in his own country of Germany, and Wernher von Braun, who was a key figure in the post-war United States space programme, and master-minded the launch of America's first successful artificial satellite, Explorer 1. In books such as *The Conquest of Space* (with Willy Ley) and articles in Collier's magazine (with von Braun) Bonestell made proposals which had a real influence upon the face of space research. There was, for instance, the Von Braun space station – a wheel, with the living quarters in the rim; rotating the wheel would simulate gravity. The actual space stations, built several decades later, did not follow this pattern, but at least Bonestell's drawings provided food for thought.

Looking at Bonestell's paintings today it is obvious that in some cases his interpretation was wrong; he showed the Moon's mountains as sharp and jagged, which was reasonable enough in view of the fact there is no lunar atmosphere to cause erosion. But he kept to the facts to the best of his ability, even when painting views of planets orbiting far-away stars. Remember, it was only in the 1990s that we detected the first extra-solar planets; Bonestell had died in 1986 at the advanced age of 98.

Others followed. Lucien Rudaux, a French astronomer, painted impressions of the lunar surface, and as he was himself a leading authority on all matters connected with the Moon he was able to improve on Bonestell's jagged mountains. In England, Ralph Smith came to the fore, he was an engineer, and some of his drawings of possible spacecraft have proved to be very close to the truth. He illustrated some of the early books by Arthur C. Clarke, who was rapidly becoming known as a 'visionary'. Then came David Hardy, now accepted as the successor to Bonestell; he has the advantage of being extremely well versed in astronomy – and this was also true of Paul Doherty, not only a splendid artist but also an expert planetary observer.

Paul was a close friend of mine, and I have vivid memories of the times we observed together. On one occasion, in the 1970s, we had spent several hours in my observatory at Selsey, using my 15-inch (380-mm) reflector to look at Saturn. Before the night was over Paul had produced the painting given here.

David Hardy has pointed out that different space artists use very different techniques and styles. In attempts to show celestial objects as clearly and accurately as possible, many artists do keep to realistic and representational styles, and this has been made even more true by the development of computers, which makes it possible for artists to use '3D' techniques. There is also a distinction between what may be called 'space art', covering impressions of space vehicles and astronauts, and 'astronomical art', for example a view of a specific planet as seen from its satellite.

Many artistic paintings have been decidedly useful to researchers – in imagination the artist can 'boldly go' where, as yet, the astronomers and the astronaut cannot. In our *Sky at Night* programme for February 2003, we invited viewers to send in their ideas for a future Martian base. Some of the entries are reproduced here, and – who knows? – some of the suggestions may well be taken up. A permanent base on Mars now is much less futuristic than a trip to the Moon would have seemed when Nasmyth and Carpenter wrote their book 130 years ago. Looking further ahead, we can visualize a time when Mars has been 'terraformed' – that is to say turned into a world as welcoming as the Earth.

There are paintings of violent scenes, too – the destruction caused by the impact of an asteroid, of the type which many people believe to have caused the climatic change which wiped out the dinosaurs, 65 million years ago. And artists can also show impressions of exploding stars, asteroid collisions, the meeting of two galaxies.... The scope is endless.

We must also consider a different branch of space art – allied not to proved facts, but to 'space opera' of the *Star Wars* variety. Aliens are always in the news, most of them decidedly unfriendly and modelled on the murderous Martians described by H.G. Wells in his classic *War of the Worlds.* I have never understood why aliens have such a bad press; if any travellers from afar can come here, they will be far in

▲▶ Future Martian base? *Reproduced here are more entries to our space art competition. Above left is the work of Christopher Harrison; above right is that of Jeffrey John; at right is the work of Dan Bright.*

advance of us, and will have put war behind them. Percival Lowell believed in civilized Martians who had built a planet-wide irrigation system with well-planned canals; it is a pity that the canals have been shown to be non-existent.

Many stars are attended by planets; there is no longer any doubt about this. The artist has full rein. What will be the scene on the planet of a dim red dwarf star, or a planet moving round a star which brightens and fades over a short period? There may be planets in binary systems, so that the landscapes will be lit by suns of different colours. So far as other races are concerned – well, some may be like us, others not even remotely humanoid. We cannot yet decide, possibly the artists can give us some useful tips. Their work is important. The International Association of Astronomical Artists (IAAA) was formed in 1981; it holds exhibitions and workshops in some of Earth's most 'alien' places, such as Iceland, Hawaii and Yellowstone

Park. It also meets at NASA establishments, and its gatherings are always well attended. Moreover, some astronauts and cosmonauts, such as Al Bean and Alexei Leonov, are themselves very skilled and recognized artists. David Hardy is, of course, one of the leading lights of the IAAA; he was President from 1996 to 2000, and is currently Vice-President for Europe.

There are some people, such as myself, who are hopelessly inartistic, and to whom perspective is a closed book. My mother was very different – a talented artist, who delighted in producing pleasant-looking if anatomically improbably beings she called 'bogeys'. Here is one of my favourites.

▲ **Pipes of Uranus?** One of my mother's paintings, it shows music and dancing on a green-tinged planet.

17 • THE UNSEEN UNIVERSE

Fritz Zwicky (1898–1974) was one of the most extraordinary characters of modern astronomy. He was born in Bulgaria, but took Swiss citizenship, and spent most of his career in California. To say that he was 'difficult' is to put it mildly. He used to show off his immense physical strength by doing handstands in the dining hall during lunchtime; he once ordered his assistant to fire bullets through the open slit of the observatory dome to check whether it improved the seeing (it didn't); he referred to his colleagues as 'spherical bastards' because he said that they were bastards no matter from which direction you looked at them, and anyone that disagreed with his views was automatically classed as a mortal enemy. But of his brilliance there was no doubt at all, and it was Zwicky who made one of the most important of all astronomical discoveries: he realized that, to our eyes, well over half the material in the universe is absolutely invisible.

▲ **Fritz Zwicky,** who is perhaps best known for his work on supernovae, galaxy clusters and dark matter.

Galaxies occur in clusters (our own Galaxy is a member of the Local Group of systems, which includes the Andromeda and Triangulum spirals and a couple of dozen dwarf galaxies). One major cluster, including 2000 galaxy systems, lies in the constellation of Coma Berenices; it is around 280,000,000 light-years away. Zwicky measured the motions of the Coma cluster galaxies, and found that they were moving too fast to keep the system stable; they had too much kinetic energy. Something was 'glueing' them together. The amount of visible matter in the cluster was hopelessly inadequate, so that there had to be a great deal of 'dark matter' whose nature was unknown (Zwicky actually called it 'missing mass'). What was it? Zwicky did not know, but it was certainly real.

Evidence came later from spiral galaxies, including our Milky Way system. Let's look first at the Solar System – the planets orbiting the Sun.

According to Kepler's Laws, orbital velocity decreases with increasing distance; Mercury moves faster than Venus, Venus faster than the Earth, Earth faster than Mars, and so on – because most of the mass in the Solar System is concentrated in the Sun. The Sun is orbiting the centre of the Galaxy, and takes around 225,000,000 years to make one circuit. A star further away from the centre ought to move more slowly – but this is not so. As in other spirals, the outer regions orbit just as quickly as those closer-in. This means that the main mass is not concentrated in the centre; dark matter is spread throughout the entire galaxy.

But have we any clue as to the nature of this dark matter, which is now believed to account for about 90 per cent of the total mass of the universe? We cannot see it, because it sends out no radiation at all. We must begin by 'weighing' a typical cluster of galaxies and then adding up the contributions from all the galaxy clusters we know – but there may be (and probably is) a vast amount of dark matter between the clusters. In 2002 a team of British and Australian astronomers completed the largest galaxy survey to date, the so-called 2dF survey of 220 million galaxies measured using a new spectrograph on the Anglo-Australian Telescope at Siding Spring, New South Wales. (2dF stands for 'two-degree field'.) The survey showed a characteristic distortion pattern due to galaxies flowing towards concentrations of dark matter. The pattern made it possible to measure the amount of dark matter distributed between the galaxies.

Another method depends upon measuring small irregularities in the 'microwave background radiation', the last manifestation of the Big Bang, when the universe as we know it came into being 13.7 thousand million years ago. The radiation dates back to a mere 200,000 years after the Big Bang; represent the age of the universe by one year and the radiation we are measuring was written on 1 January. The radiation was discovered in 1966, and at first seemed strangely uniform, but in 1993 the Cosmic Background Explorer satellite (COBE) detected patchiness – to the great relief of theorists! This patchiness gives us a fossil record of the galaxies we see today, and the sizes of the fossils gives us a clue to the amount of dark matter in the universe. More recently the 'Boomerang' balloon experiment over Antarctica has provided extra information. The amount of dark matter is now given as just under 90 per cent of the total cosmic mass.

◀ *The Coma cluster of galaxies,* imaged by the 4-m Mayall Telescope at Kitt Peak.

▶ *Boomerang, which stands for Balloon Observations of Millimetre Extra-Galactic Radiation and Geophysics, measured temperature variations in the cosmic microwave background and compared the results (top) to computer models (bottom). The model that most closely resembled the results was that of a flat universe.*

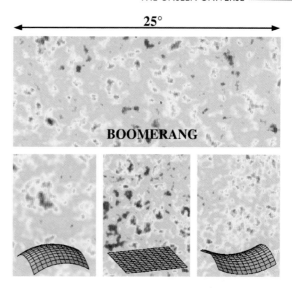

25°

BOOMERANG

But what is it?

We do not know – but we do know what it is not. It is not ordinary matter of the type making up the Sun, the planets, you and me. The Big Bang may be likened to a nuclear furnace. About three minutes after the Bang, the first chemical elements were manufactured in this furnace: hydrogen, the simplest element in the periodic table, and helium, the next simplest. One of the great successes of the Big Bang theory is that it predicts exactly the right amounts of these elements found in gas clouds present when the universe was still very young, and the first galaxies were starting to form. The amounts of hydrogen and helium measured show that the total mass of ordinary matter can be no more than around 10 to 15 per cent of the total mass of the universe. The remainder must be something else – something exotic.

For many years, particle physicists have been predicting the existence of exotic elementary particles that are totally unlike the atoms we know. They have not yet been detected, because they interact only weakly with other matter – which is why they are 'dark'. Collectively, they are referred to as CDM, Cold Dark Matter.

To detect it, the obvious course is to design a machine capable of doing so, but this is easier said than done, though energetic efforts are being made. The experimental equipment has to be shielded from cosmic rays, and must therefore be taken well below the Earth's surface – such as Boulby Mine in England. Up to now no positive results have been obtained.

A different kind of 'experiment' has been tried. We can try to create many universes by computer studies, each with a different amount of matter, and can then check to see which model matches our actual universe. However, the simulations show that universes without any dark matter, or universes in which dark matter is anything but CDM, fail miserably; either they make no galaxies, or they make galaxies that are much too big or much too small. Only CDM simulations produce really credible universes.

So are we any further in deciding upon the nature of dark matter? Is the missing mass locked up in Black Holes? Are there vast numbers of undetectable low-mass stars? Do the amazingly abundant neutrinos have any mass? All these explanations are unsatisfactory; we come back to exotic matter which we cannot detect in any way. One day we may understand it – but not yet.

18 • MERCURY CROSSING

On 7 May 2003, I watched an interesting event from my observatory at Selsey: the little planet Mercury crossed the face of the Sun, and I had a perfect view of it.

Mercury and Venus are known as the interior planets, because they are closer to the Sun than we are. On average Mercury is 58 million km out (36 million miles) and Venus 108 million km (67 million miles), and at times when there is an exact lining-up, a transit is the result. With Venus this does not happen very often; there were two transits in the 19th century, in 1874 and 1882, and none in the 20th. Mercury is more co-operative; there were 14 transits between 1900 and 2000. I had never been successful in seeing one, however, so I looked forward to this month's event. Transits do not occur every time Mercury is 'new', because the orbit is tilted at an angle of just over 7 degrees. Transits can occur only in May or November.

Ptolemy, last of the great astronomers of Classical times, realized that Mercury and Venus could transit, but the first prediction of a transit was made by Johannes Kepler, who found that a transit would occur on 7 November 1631. It did – and was observed by the French astronomer Pierre Gassendi. It was Edmond Halley who realized that transits could be used to measure the length of the astronomical unit, or Earth–Sun distance. The start and end of a transit (known as ingress and egress) will occur at slightly different times as seen from different locations on the Earth's surface, because the diameter of the Earth is not negligible compared with the distances of Mercury and Venus. Exact timings make it possible to work out the distance of the planet – and Kepler's Laws can then

▼ **The transit of Mercury,** on 7 May 2003, seen in a time-lapse *image from SOHO that clearly shows its progress across the face of the Sun.*

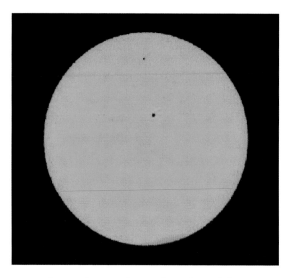

◄ **Mercury transit,** imaged by John Fletcher. Mercury, the dark speck to the upper left of centre, is not to be confused with the large sunspot on the centre of the Sun's disk.

be used to work out the value of the astronomical unit. But Halley also realized that Mercury would look so tiny that precise measurements would be difficult; Venus would be better. In fact transits of Mercury, interesting though they are, could never be regarded as really important.

During the 2003 transit, the apparent diameter of Mercury was 12 seconds of arc – that is to say, 1/158 the apparent diameter of the Sun. It could not be seen with the naked eye; a telescope was essential – and before the transit I was careful to give the usual warning about looking directly at the Sun. Of course proper filters are safe enough, but I preferred to use my 5-inch (125-mm) Cooke refractor to project the Sun's disk on to a screen fixed behind the eyepiece.

On the credit side, a transit of Mercury is a leisurely affair. On this occasion ingress fell at 05.13 GMT, and egress was not due until 10.32 GMT; there was no mad rush of the type associated with a total eclipse of the Sun.

The first thing that struck me, when Mercury had passed on to the solar disk, was that the planet was obviously blacker than any part of a fairly large sunspot which happened to be on view at the time. This was only to be expected. The unilluminated hemisphere of Mercury is indeed jet-black, because the Mercurian atmosphere is far too varied to cause any twilight effects, whereas a sunspot is certainly not genuinely dark – if it could be seen shining on its own, its surface brightness would be greater than that of an arc-lamp. It looks dark only in contrast against the surrounding photosphere. And Mercury was indeed small – smaller than, subconsciously, I had anticipated. It moved slowly but steadily onward, in the northern part of the Sun's disk, and it was fascinating to follow its progress.

Secondly, there was no 'Black Drop' effect. When Venus enters transit, it seems to draw a strip of blackness after it; when this disappears, ingress has already occurred – with the result that very precise timing is impossible. It was this Black Drop that effectively ruled out Halley's method for measuring the astronomical unit. It is due to Venus' atmosphere, and nothing comparable would be expected with airless Mercury. Using as high a magnification as I could I watched closely for any sign of a Black Drop, and predictably the results were completely negative. Also, throughout the transit the little disk of Mercury was absolutely sharp. The trace of atmosphere could produce no visible effects at all.

Another investigation made at past transits involved the search for a Mercurian satellite. This would have been very difficult even if a satellite existed, but we are now sure that Mercury, like Venus, is a solitary traveller in space.

What else? Well, during past transits there have been occasional reports of bright specks on the disk, and other phenomena that are certainly to be dismissed as nothing more significant than optical effects. Observations have been made at X-ray wavelengths and in hydrogen light, but I was content to do no more than watch the little planet as it made its way from one limb of the Sun to the other. I knew that I was not likely to have another chance. The next transit accessible from England is that of 2016, and frankly I do not expect to live to the advanced age of 93. But others will, so the table below gives the transits due in our century.

Happy viewing!

8 November 2006, mid-transit	21.42 GMT
9 May 2016	14.59
11 November 2019	15.21
13 November 2032	08.55
7 November 2039	08.48
7 May 2049	14.26
9 November 2052	02.32
18 May 2062	21.39
11 November 2065	20.08
14 November 2078	13.44
7 November 2085	13.37
8 May 2095	21.09

19 • HIGHLAND RING

There is no doubt in my mind that a total eclipse of the Sun is the most magnificent sight in all Nature. As the last sliver of the bright disk vanishes, you see the 'diamond ring', the red prominences, and above all the pearly corona. But from any particular point on the Earth's surface, total eclipses are uncommon. There have been two in England during my lifetime. The first was in 1927, when the track crossed Wales and Yorkshire; I missed it, because I was only 4 years old and lived in Bognor Regis. The second was in 1999, from Cornwall, which I missed because of a cloudy sky. But there are also annular eclipses, which are much less spectacular but are nevertheless well worth watching.

It is sheer chance that as seen from Earth the Sun and the Moon look almost the same size; the Sun's diameter is 400 times that of the Moon, but the Sun is also 400 times further away. However, the Moon's path round the Earth is not circular, and the distance from us ranges between under 370,000 km (230,000 miles) at apogee to just over 400,000 km (250,000 miles) at perigee. When the lining-up occurs with the Moon near perigee, the lunar disk is not large enough to cover the whole of the Sun, and a ring of sunlight is left showing

▼ **Annular eclipse party.** *Waiting on the Scottish coast are Brian May and Iain Nicolson to my right, and Steve Wainwright to my far left.*

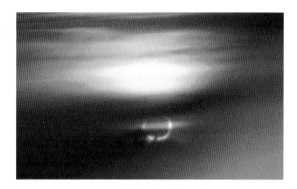

► **Annularity.** *The bright ring of sunlight surrounded the Moon's disk can be seem clearly, despite the clouds, in these two photographs.*

round the dark Moon. This is an annular eclipse (Latin *annulus*, a ring). From Britain, there was an annular eclipse on 8 April 1921 and another was due on 31 May 2003. Clearly *The Sky at Night* had to be there, but the eclipse was due a few minutes after sunrise, so the altitude would be very low. A better view was expected from Iceland, but the BBC could not afford to send a camera there (!) and so we repaired to Talmain in Sutherland, between Lossiemouth and Cape Durness. With me were Iain Nicolson, Dr Steve Wainwright and Dr Brian May. (How many people realize that the great guitarist is also a highly qualified astronomer?) We set up our equipment, and hoped for the best.

It was certainly a lovely scene. Rabbit Island lay in the foreground; in the distance, during daylight, the Orkneys could be dimly seen. What worried us was the amount of low cloud and mist, just in the area where we did not want it. There were clear areas too, however, and the outlook did not seem to be hopeless. By 3 a.m. British Summer Time we were ready. Sunrise was due at Talmain at soon after 4 a.m., by which time the partial phase of the eclipse was well under way. Annularity was timed for 04.42, and would last for three minutes.

Our programmes were different. Brian and Steve had infrared cameras, so that if thin clouds hid the Sun they might still be able to

▲ *After annularity.* By *well above the horizon –*
the time I took this photo- *and above the clouds. The*
graph, the Sun had risen *crescent shape is clear.*

image it. Iain had ordinary photographic equipment. I had very little; my main role was to give a commentary.

By 04.30 there was quite a crowd of people (as we learned later most onlookers had gone to Durness, and the Press had expected us there). I did several radio and TV interviews; so did Brian – making it clear that on this occasion he was a scientist and not a musician. All of us kept a close eye on the eastern sky. The clouds were still there, but a dim rosy glow betrayed the position of the Sun. Neither Brian nor Steve had any images in infrared; I thought that we were doomed to be frustrated.

Two minutes to annularity – and Steve called out, 'I can see it!' On his screen there was a faint but very evident picture of the Sun – or rather a ring of the Sun round the dark mass of the Moon. Brian, a few metres away, recorded it at the same time. Visually, the Sun was still hidden, but I took a photograph more in hope than in expecta-tion. On Steve's screen we could see Baily's Beads, caused by the sunlight streaming through valleys on the Moon's limb; the annular-ity was over – and at that instant the clouds broke, and we could see the still mainly eclipsed Sun.

During annularity the light had had rather an eerie quality, reminding me of the 1999 total eclipse from Falmouth in Cornwall; then we had not seen the Sun at all, but it was strange all the same. Now, at Talmain, the light was coming back, and there was displayed the crescent Sun. I remember Iain commenting that the apparent diameter of the Moon was obviously less than that of the Sun; in fact only 94 per cent of the Sun was covered even at maximum obscuration.

I finished my commentary, and called in Steve, Iain and Brian as we watched the waning partial phase. The clouds were still there, but we no longer cared about them; we may not have had a perfect view of the 'Highland Ring' but we had seen enough to satisfy us and to put together a proper programme. Talmain had treated us well. Later I found that the conditions at Durness had been much the same; Shetland was patchy. The very best views had been obtained from Iceland, well north of Reykjavík, where John Mason and Michael Maunder had stationed themselves.

Next time? Well, future British annulars are due on 23 July 2093 (northern England, southern Scotland, Northern Ireland), 12 April 2173 (just misses the Hebrides), 27 September 2220 (western Ireland, Land's End) and 27 March 2294 (Orkney, Shetland). I fear I will not see any of these, but I hope the astronomers of the time will enjoy them. At least I have pleasant memories of 31 May 2003.

20 • GAMMA-RAY BURSTERS

Bodies in the sky send out radiations all through the extent of the electromagnetic spectrum, from very long radio waves through to the ultra-short gamma rays. Visible light makes up only a very small part of the whole spectrum, but many of the radiations from space are blocked by the Earth's atmosphere – which is fortunate for us, since many of the rays are harmful, and without atmospheric protection you and I would not be here. Gamma rays are absorbed in the upper air, and only the most energetic can reach ground level, so virtually all research has to be carried out by using rockets or satellites.

The first detection of high-energy gamma rays from space was achieved by the Explorer XI satellite, in 1961. There followed many specially designed satellites, such as CosB (launched 1975) and the immensely successful CGRO or Compton Gamma Ray Observatory (1991), named after the great pioneer of this research, Arthur Holly Compton. GRO was deliberately crashed into the sea in June 2000, as it had come to the end of its career and the equipment was starting to fail.

Of special interest are gamma-ray bursters (GRBs). The first of these was detected in 1967 by a satellite (Vela), and hundreds have

◀ ▼ *Gamma-ray burster.* *Discovered in March 1997, GRB 970228 actually took place on 28 February that year. The Hubble Space Telescope obtained this false-colour image in September 1997, by which time the GRB had faded to 1/500 of its original brightness.*

▲ **GRB 990510**, *imaged by the Antu unit of the Very Large Telescope. The afterglow of the GRB is clearly visible in the boxed area in the lefthand image when compared with the same area in the righthand image, which was taken before the GRB.*

been recorded since. They last from a few seconds to a few minutes, and may appear from any direction at any moment. They are immensely powerful and remote; GRB 000131 (31 January 2001) seems to have been about 11,000 million light-years away, and to have become a million million times more luminous than the Sun. But what exactly are GRBs?

For a long time nobody had any real idea. It was widely believed that they were phenomena of our Galaxy, a few hundreds of light-years away, and that they were probably caused by a head-on collision between two neutron stars. Then, in 1997, the distance of a GRB was measured for the first time, and the 'local' theory had to be jettisoned. The initial problem was that the detectors could provide only rough positions for GRBs – it was rather like trying to identify the source of a sound when you can't see it. Then, in 1996, came a Dutch–Italian satellite, BeppoSAX, which was able to observe GRBs both in gamma rays and with X-ray telescopes. (BeppoSAX lasted for years, finally descending into the sea in May 2003, its work done.) X-ray telescopes produce a fuzzy picture of the area they are looking at, and this was good enough to give a much more precise position for the bursts. It then became possible to use large ground-based telescopes, with which were found fading points of light whose distances could be determined by the red Doppler shifts in their spectra. Over 140 GRBs have now had their distances measured in this way.

Apparently GRBs occur in 'starburst' galaxies, where new stars are being formed rapidly (M82 is one such galaxy). If large numbers of stars are formed, some of these will be particularly massive, and will live for a relatively short time before running out of 'fuel' and collapsing to form neutron stars or black holes. In many cases this sort of collapse leads to a supernova explosion. In rare circumstances such a collapse may also produce a very high-energy jet of matter, and if we happen to be looking straight down the jet we see a gamma-

◀ **GRB 050509b,** detected by the Swift satellite on 9 May 2005. The XRT instrument pinpointed the GRB to within a small area adjacent to a large, distant elliptical galaxy, as shown in this image obtained by the 10-metre telescope at the W.M. Keck Observatory. S1–4 are possible optical counterparts of the GRB, but none has been confirmed.

ray burst. A prediction of this picture is that we ought to see something like a supernova accompanying every GRB, but it was only in April 2003 that the first definite detection of a supernova with a GRB was recorded. The main trouble is, of course, that the light from the supernova is drowned by the GRB. Many problems remain. For instance, some GRBs seem to be much shorter-lived than the average, lasting for less than a second, so that they may well be in a completely different class.

Because they are so powerful, GRBs can be seen over great distances, and some to be found within the next few years may well prove to be the most remote objects ever observed; they may tell us something about the earliest structures in the universe. At its peak, the most luminous GRB we have seen shone with the intensity of more than a million supernovae – about a thousand times brighter than the most luminous quasars. Since GRBs are so brief, you have to begin observing one within minutes or, at most, hours to catch it while it is still very powerful. One GRB, seen in 1999, reached magnitude 9, but more generally they are between magnitudes 13 and 18 when picked up. When a satellite detects a GRB, it promptly radios the position to the ground, and the data are at once distributed over the internet. A few telescopes are designed to respond immediately to these alerts, and they begin observing automatically. With most large professional telescopes, however, the request to override the normal programme has to come from a human, so many astronomers around the world receive these alerts by e-mail and mobile phone messages (at any time of the day or night) and then contact telescopes to which they have priority access.

Initially, the aim is to measure the light variations at all wavelengths between X-ray and radio. Optical data are particularly important because of the relative ease of measuring red shifts. Small telescopes can be used for brighter GRBs (say above magnitude 18), and amateurs can join in; for example John Fletcher, of Gloucester, obtains valuable results with a CCD on his modest 10-inch reflector. Amateurs are becoming more and more proficient.

In November 2004 a new satellite, Swift, was launched. A joint effort by Britain, Italy and the USA, it should be able to locate at least 10 GRBs a month and send out positional data in a matter of seconds. This makes it possible to do some really good statistical analysis – and many of the GRBs will no doubt prove to be particularly interesting.

GRBs have been known for less than half a century, but valuable data have been obtained. We do not yet fully understand them, but at least we are now sure that they are the most colossal explosions ever recorded by mankind.

— 21 • THE VERY EARLY UNIVERSE —

How old is the universe? For many years astronomers could not agree. It was generally accepted that everything – space, time, matter – was created in a 'Big Bang', but how long ago did this happen – in other words, how old is the universe? Estimates ranged between 10 thousand million years and 20 thousand million years. Now, at last, we are confident that we can give a correct value: 13.7 thousand million years. Space, time and matter were all created at that moment.

There are various types of matter. There is ordinary matter, of the type with which we are so familiar; and there is dark matter, which does not shine at all. There is about seven times more dark matter than ordinary matter: though we are not sure what it is, we have every reason to suspect that it is made up of elementary subatomic particles. It generates the lion's share of the gravitational force that regulates the evolution of the universe.

In early times the universe contained a large number of quasars, enormously energetic and powered by supermassive black holes in their centres. The all-pervading gas – mainly clouds of hydrogen – is detected in an ingenious manner. A cloud may intercept the light from a background quasar. When this happens the light of a particular wavelength is absorbed by the hydrogen cloud, blocking out part of the quasar's light. It is also possible to deduce how much obscuring material is present.

By about three thousand million years after the Big Bang, we find that the stars have already formed into galaxies. The existence of galaxies at that epoch was first established by using the Hubble Space Telescope, which was pointed at a particular small patch of the sky (about the area of a postage stamp) and kept there for many orbits. This so-called Hubble Deep Field revolutionized our ideas about the early universe. It revealed galaxies so remote that their light has taken most of the lifetime of the universe to reach us. These images tell us what the early galaxies were like – going back to a time when the universe was only 5 per cent of its present age.

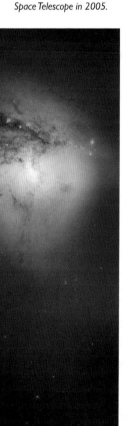

▼ **The Whirlpool Galaxy (M51, NGC 5194)**. *It appears to be interacting with the smaller galaxy NGC 5195. This image was taken by Hubble Space Telescope in 2005.*

We are still trying to decide how many galaxies existed then and, more specifically, how many of the stars we see today had already been formed. The idea that we might be able to track the entire history of star formation was proposed near the end of the 20th century, when the first galaxies caught in the act of forming stars for the first time – the so-called Lyman-break galaxies – were discovered with the William Herschel Telescope on La Palma. Alas, these prospects were soon dashed because of a second major discovery, made with the James Clerk Maxwell Telescope in Hawaii. It was equipped with a new detector, SCUBA, which is able to measure light of sub-millimetre wavelengths (typical of the wavelengths used for radio transmissions on Earth). SCUBA detected many young galaxies that were producing stars, but shrouds of dust obscured the regions where this star formation was going on, making it very difficult to find out how many stars

◀ **The end of the Dark Ages,** as captured by the Hubble Space Telescope. Astronomers believe that the faint objects marked with arrows are young star-forming galaxies seen just a thousand million years after the Big Bang, as the so-called Dark Ages were coming to an end.

were actually being produced. Even today the situation is still unclear.

In 1992 the Cosmic Background Explorer satellite (COBE) discovered small ripples in the radiation left over from the Big Bang – the Cosmic Microwave Background (CMB). Theorists had predicated these ripples had to be there – otherwise, no galaxies could have formed. According to the theory, when the universe was born it was not completely smooth (that is to say, uniform from place to place), but contained small irregularities – patches where the temperature and density were slightly different from the average. These irregularities were caused by subatomic or 'quantum' processes which had occurred when the entire universe was smaller than a pinhead. As the universe expanded, the irregularities became more and more pronounced, and would be expected to leave imprints in the CMB – very small, but sufficient to be detected with very sensitive equipment. In 2003 another satellite, WMAP, studied the CMB in more detail than COBE had been able to do – and the details of the small fluctuations were in exact agreement with the theory.

The significance of the small imprints in the CMB is that they are nothing more nor less than the progenitors of the galaxies, clusters of galaxies, and superclusters that we see today. This continued until they collapsed to form objects made up of a mixture of dark matter and ordinary matter. The dark matter was dominant, and settled into a clump – known as the dark matter halo – inside which ordinary matter formed a disk, which then began to fragment into stars. This, then, seems to be how galaxies were born.

Soon after the emission of the background radiation, the growth of galaxies was interrupted by the appearance of the first quasars and proto-galaxies, which generated enough radiation to ionize the ordinary matter, heating it up and preventing it from falling into the dark matter halo. This period, lasting around a thousand million years, has been called the Dark Age – an appropriate name, because the universe was indeed dark and opaque. The dark haloes continued to grow, but no stars formed within them until the gas had had sufficient time to cool down. At that point the 'Renaissance' occurred: the universe suddenly blazed with light as millions upon millions of stars formed – the stars we see today. The process of galaxy formation was well under way.

Powerful computer simulations can be used to follow the growth of galaxies. They show that the first dark matter haloes, which form together with their associated galaxies, are small fragments. The fragments are attracted to each other by gravity, and collide, fusing to make larger haloes and larger galaxies. During a merger, the dark matter haloes simply coalesce, but the galaxies within them can be distorted or disrupted. Graceful spirals may produce large elliptical systems. Theory tells us, and computer simulations confirm, that

▶ **18 small young galaxies,** in this HST image, are at a distance of 11 thousand million light-years, when the universe was 16 per cent of its present age. They are small enough and close enough together to be the building blocks of today's galaxies.

▼ **The CMB,** imaged by the Wilkinson Microwave Anisotropy Probe. The different colours represent temperature fluctuations.

galaxies were assembled gradually by putting together fragments – a sort of hierarchical process that began two or three thousand million years after the Big Bang. Mergers make large galaxies, and we can see evidence of this, for example with the Tadpole galaxy, which is luring a smaller galaxy towards it; the victim will be absorbed in what is often called cosmic cannibalism. The Tadpole is also associated with young globular clusters, which are being formed because of the encounter. The origin of globular clusters is not well understood, but our Galaxy contains about a hundred of them. The implication is that our Galaxy swallowed up another galaxy a few thousand million years ago. In the Tadpole we have a glimpse of what our galaxy may have been like in its infancy.

Large galaxies of comparable size occasionally run into one another, as the so-called Mice galaxies are doing. They are drawn together by gravity, and after performing a sort of cosmic dance they merge, triggering violent star formation together with X-ray and ultraviolet emissions. Such encounters leading to massive mergers are rarer than encounters leading to minor mergers, but on

▼ *The Mice galaxies.*
Two spiral galaxies are in
the process of merging to
form a single giant galaxy.

▶ *The Tadpole.* Its long tail
and bright star-forming
regions result from a collision
with a smaller galaxy.

average every galaxy may expect at least one during its lifetime. Our Galaxy will eventually collide with the Andromeda Spiral, which is at present over two million light-years away; the force of gravity is inexorably pulling them together. Luckily the collision will not happen for thousands of millions of years, so that there is no immediate cause for alarm. If the Tadpole shows us what our Galaxy may have been like thousands of millions of years ago, the Mice take us equally far into the future.

We have only a fuzzy sketch of how our universe became structured into galaxies, but we believe that the origin of galaxies goes back to the origin of the universe itself, when galaxies were seeded by processes occurring on subatomic scales. We see these seeds three hundred thousand years later as tiny ripples in the microwave radiation, and then again, three thousand million years later, in the Hubble Deep Field. The future is tremendously exciting; new surveys with new instruments will surely lead us to a coherent theory of galaxy formation, and hence of the evolution of the universe over the past 13 thousand million years.

— 22 • SMART WAY TO THE MOON —

On 27 September 2003, Europe's SMART-1 probe was sent on its way to the Moon. It was in no hurry. Previous missions had taken only days to reach lunar orbit, whereas SMART-1 would take months, and was not scheduled to arrive until December 2004. It was using a new method of propulsion – an ion engine.

The launch of a conventional lunar probe is striking. I well remember watching the ascent of Apollo 17, the last manned mission, in December 1972 – how long ago that seems! Of course I was standing well back, as we all were, but it was a night launch, which made it all the more impressive. The light from the rocket's exhaust was brilliant, and then came the 'wall of sound' as the huge Saturn V thundered into space carrying the final Apollo astronauts. As we know, the mission was a total success, and Gene Cernan and geologist Harrison Schmitt walked happily around on the lunar surface. Plans for Apollos 18 to 21 had been made, but were never implemented. Remember, there was no rescue provision, and we must admit that Apollo had done everything it was capable of doing. If the flights had

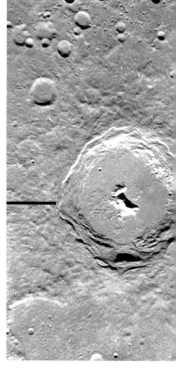

▲ **The crater Pythagoras,** with its central peaks and terraced walls, is seen in great detail in this mosaic of images returned by SMART-1. The spacecraft entered lunar orbit in November 2004.

continued they would have added only a limited amount of information, and there was always the danger that something would go horribly wrong. Since then there have been unmanned missions, but there are still problems to be solved. In particular, it has been claimed that some of the polar craters, whose floors are always in shadow, are covered with ice. I am a total sceptic here, but we must wait and see.

Missions to Mars are also being planned, and indeed several unmanned probes have been launched this year (Europe's Mars Express, and the American rockets carrying the rovers Spirit and Opportunity). But chemical propellants have their drawbacks, and there are many advantages in using atomic power. This is what SMART-1 is doing.

The technique has been used twice before, by NASA's Deep Space 1 and ESA's experimental communications satellite Artemis, but not in the same form, so that SMART-1 is breaking new ground. It is the first of the Small Missions for Advanced Research and Technology – hence its name. Instead of using chemical propellants, it depends upon atoms of xenon, an inert gas which is found in very small quantities in the Earth's atmosphere.

The trick is to use electric power from the satellite's solar panels to accelerate charged atoms of xenon through the engine. The thrust is tiny – in technical terms 0.07 Newton, which is about the weight of

► *SMART-1's ion engine,* an artist's impression. It gets its thrust from the ionization of xenon gas.

▼ *SMART-1 in lunar orbit,* an artist's impression. The mission will return data on the Moon's geology, mineralogy, morphology and topography, as well as testing the solar-powered ion drive.

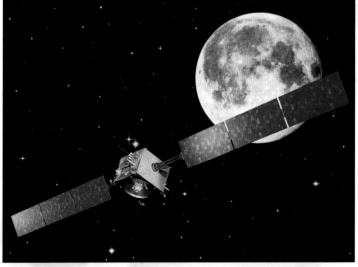

◀ *Autumn 2004, and SMART-1 is approaching the Moon. It entered lunar orbit on 15 November 2004 and is the first European spacecraft to reach and orbit the Moon.*

28 October 2004
~ 600 000 km from the Moon

12 November 2004
~ 60 000 km from the Moon

a postcard, so that SMART certainly does not roar into space. As soon as it has been released from its launching vehicle, it depends entirely on the thrust of the xenon ions, drawn from 82 kg of the gas being carried. The ion motor can fire over a very long period, and the effects are cumulative.

The first manoeuvre is to spiral outward from the Earth, moving in an elliptical path. When it has receded to about 200,000 km (125,000 miles) from the Earth, it will start to come under the influence of the Moon. Gravity-assist manoeuvres will be performed during fly-by passes of the Moon in December 2004 and January and February 2005, and eventually, in March 2005, it is due to enter lunar orbit and begin its scientific programme. It carries many instruments, and will – we hope! – be able to send back data about the Moon's mineralogy as well as continuing the search for ice (which I am confident will be unsuccessful). Another advantage of the ion drive is that SMART-1 can be 'steered' in a way that a chemical rocket cannot. Its lunar orbit will take it above the poles, and the distance from the Moon's surface will range between 800 km (500 miles) and around 16,000 km (10,000 miles).

For longer journeys, beyond the Earth–Moon system, the ion motor will eventually be preferred, and it has been claimed that given enough time, the ion tortoise will overtake the chemical hare! So a great deal hinges upon the performance of SMART-1. If it works as well as its makers hope, then it will encourage those engineers who are planning flights to Mercury and to the outer planets.

23 • THE GALILEO MISSION

On 23 September 2003, one of NASA's most successful projects came to an end. After an eventful and immensely profitable career, the Galileo spacecraft was deliberately crashed into Jupiter – a classic example of cosmic suicide.

Jupiter is the giant of the Solar System, and is more massive than all the other planets combined. It has a hot silicate core, overlaid by a layer of metallic hydrogen; above this comes a layer of molecular hydrogen, and then the cloudy atmosphere, composed mainly of hydrogen and hydrogen compounds but with a good deal of helium. There is a very powerful magnetic field, and an extensive family of satellites. The so-called Galilean satellites are of planetary size; Io is slightly larger than our Moon, Europa slightly smaller, and Ganymede and Callisto much larger. Indeed, Ganymede, with a diameter of over 4800 km (3000 miles), is larger than the planet Mercury, though less massive. All the other satellites are small; by now over 60 are known, though some of them are almost certainly captured asteroids. Of the Galileans, Ganymede and Callisto are icy and cratered, Europa icy and smooth, and Io wildly volcanic, with major eruptions in process all the time.

The Galileo mission was nothing if not ambitious. It was scheduled to study the Jovian atmosphere and magnetosphere, and to obtain close-range information from the major satellites. The major craft carried an entry probe to be separated and crashed into Jupiter so that it would continue to transmit data until being destroyed. Despite the almost immediate failure of the high-gain antenna, the entire programme was carried through. The route to Jupiter was not straightforward; Galileo used the 'gravity assist' technique, and had to make two close passes of Earth and one of Venus before moving out to its main target. In fact, it may be said that the route was rather like driving from Brighton to Bognor Regis by way of Manchester!

It may be helpful to summarize Galileo's career:

1989 October 18	Launch from Cape Canaveral, by the space shuttle Atlantis.
1990 February 10	First Venus fly-by.
1990 December 8	First Earth fly-by.
1991 October 29	Fly-by of asteroid Gaspra.
1992 December 8	Second Earth fly-by.
1993 August 28	Fly-by of asteroid Ida.
1994 July	Images obtained of the impact of Comet Shoemaker–Levy 9 on Jupiter.
1995 July 13	Arrival at Jupiter; entered orbit. Entry probe deployed. 147 days until impact on Jupiter.
1995 December 7	Entry probe crashed into Jupiter.
1995–2003	35 orbits of Jupiter made; encounters with all the Galilean satellites as well as the small inner satellites such as Amalthea.
2002 January 17	Cameras deactivated (because of radiation damage).
2003 September 21	Official end of the mission.
2003 September 23	Crashed into Jupiter, and destroyed.

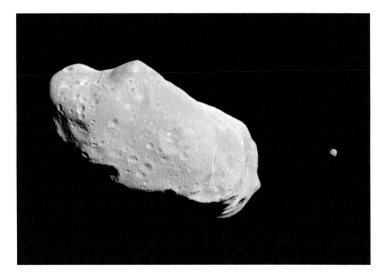

▲ **The asteroid Ida,** and its small moon Dactyl, imaged by Galileo on 28 August 1993. This was the first proof that natural satellites of asteroids exist.

The total distance travelled from launch to impact was 4,631,778,00 km (2,878,053,600 miles). About 14,000 pictures were returned. The trajectory was officially called VEEGA (Venus–Earth–Earth–Gravity Assist).

The images of the asteroids Gaspra and Ida were interesting; both these tiny worlds were irregular and cratered, and Ida was found to have a mile-wide satellite, Dactyl. Galileo was ideally placed to record the comet crash into Jupiter, whereas Earth-based observers had to wait until the planet's rotation brought the impact site into view (admittedly this did not take long; I was using the 26-inch refractor at Herstmonceux in Sussex, and the sight was truly amazing). The results of the entry proper were rather unexpected, since the vehicle encountered little cloud and few vapours; we now know that this was because it plunged into one of the most prominent dark patches near the Jovian equator. Almost cloud-free, these dark patches are known as 'hot spots' because infrared images show heat from well inside the globe radiating out from these deep, clear weather systems. It was also found that the atmospheric composition was similar to that of the Sun, apart from having two to three times more hydrogen and helium. This argues against the theory that Jupiter formed by the accretion of numerous icy mini-planets; it is more consistent with direct collapse from the gas cloud from which the Solar System developed. Data were received for 58 minutes after impact, by which time the probe was around 100 miles below the cloud-tops.

Studies of Jupiter's atmosphere and magnetosphere from the orbiter proved to be most illuminating. Windspeeds of over 640 km/h (400 mph) were detected, and an intense new radiation belt was discovered around 50,000 km (31,000 miles) above the cloud-tops. The obscure Jovian rings were found to be formed from dust hurled up by meteoric impacts on the small inner satellites, and could not be more unlike the glorious icy rings of Saturn. Many mysteries remain, however; for example what is the cause of the colour of the Great Red Spot, a huge anti-cyclone full of thick clouds? Sulphur compounds have been suggested, but we do not really know.

Of the Galileans, Io was the most spectacular, with its fiercely hot, sulphur volcanoes. Ganymede was found to have its own magnetic field – a magnetosphere within a magnetosphere! But it was the icy Europa that caused the greatest interest. Its surface area was cracked, with little vertical relief, and there were few impact craters, though one, Pwyll, was reasonably prominent. It looked as though the Europan surface was carrying icebergs, and it had already been suggested that there might be an ocean of liquid water below. Undoubtedly the interior is heated by tidal flexing caused by the gravitational pulls of Jupiter and the neighbouring Galileans, though the effects are much less marked than with Io, which is of course much closer-in to the planet.

Europa sweeps through Jupiter's magnetic field. Its own field is very weak, but is found to oscillate, and the interactions seem to show that there is a liquid layer only a few kilometres below the surface. Salty water is a good candidate; it conducts electricity well, whereas ice does not. As yet the existence of an underground sea has not been

▶ *An active volcano on Io,* in an image taken by the Galileo spacecraft on 22 February 2000. The white and orange areas represent hot lava.

proved, but it does look probable. (Surprisingly, effects of the same kind were found with Callisto, which had been assumed to be totally inert. Ganymede's much stronger magnetic field makes measurements there difficult.)

What can this sunless sea be like, if it really exists? Dark, tepid, lonely beyond belief. Can there be any life there? If so, it must be very lowly – certainly nothing as advanced as a fish. We may have a better idea when we decide whether there are any living organisms in the assumed underground seas of Mars, or for that matter in Lake Vostok, deep below the ice sheet of our own Antarctica.

Yet the jury is still out, and what we do not want to do is to carry Earth contamination to Europa. We want to examine the satellite in its mint condition, so to speak, and by 2003 the dying Galileo probe could start to be a threat. Its tasks were completed; rigid control could not be maintained indefinitely. It, or parts of it, could crash into Europa. Therefore, the decision to send it to its doom was taken. The last signals were received; Galileo passed into Jupiter's shadow, and its last hours were spent in darkness. The final chapter, the entry into the Jovian clouds, was not observed. In a way it was a strange feeling; one of the leading investigators, Rosaly Lopez, commented: 'Personally I am a little

▶ **Europa's icy surface,** showing two reddish regions, which some astronomers believe were formed by ice welling up from an ocean below.

▼ **Jupiter's Great Red Spot,** shown in approximately true colours, in an image taken by Galileo on 26 June 1996.

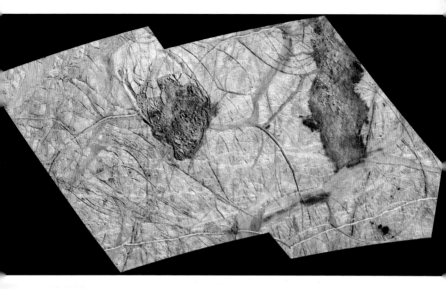

sad. I had the time of my life on Galileo, and I'm saying good-bye to an old friend.' But another leading investigator, Torrance Johnson, disagreed: 'We haven't lost a spacecraft, we've gained a new stepping stone in exploration.'

Even during the last part of Galileo's career, a new spaceprobe, Cassini, was on its way, by-passing Jupiter en route to Saturn. Others are planned, including one mission with Europa as its main target. We may or may not find life, but at least we can agree that this strange satellite is one of the most intriguing worlds in the whole of the Solar System.

24 • SPACE WANDERERS

The end of 2003 marked the return of a familiar visitor: Encke's Comet, which moves round the Sun in a period of 3.3 years – the shortest known – and has been seen at every return since 1822, apart from that of 1945, when it was very badly placed and most astronomers had other things on their minds. It was first recorded on 18 January 1786 by the French astronomer Pierre Méchain. It was seen again by Caroline Herschel, William Herschel's sister, in 1795, by Jean Pons in 1805, and by Pons again in 1818. The German astronomer Johann Encke realized that these four comets must be identical, and predicted that it would return in 1822. On 24 April of that year it was found by C.K.L. Rümker, from Australia, just where Encke had expected it. Fittingly, Encke's name was given to the comet.

It used to be brighter than it is now. It was formerly a naked-eye object when best placed, and in 1829 reached magnitude $3\frac{1}{2}$; at that time it had an obvious tail. Since then it has faded; it seldom reaches naked-eye visibility, and any tail is very short and obscure. This is not surprising. A comet loses material every time it passes perihelion, and

▼**Comet NEAT (C/2001 Q4),** *photographed by John Fletcher. First detected* *by the Near Earth Asteroid Tracking project, it has a very eccentric orbit.*

▶ **Comet Hyakutake,** *photographed by John Fletcher on 17 March 1996. It was discovered by the Japanese amateur astronomer Yuji Hyakutake on 31 January 1996.*

is bound to waste away. Encke's Comet never moves out as far as the orbit of Jupiter, and with modern instruments it can be followed all the time, even when it is near aphelion and very faint indeed.

There was one great disappointment this year. A spacecraft, Contour (*Comet Nucleus Tour*), was scheduled to fly past Encke in November, and again in June 2006 and August 2008; it would have also encountered two more short-period comets, Schwassmann–Wachmann 3 in 2004 and D'Arrest in 2008. It was launched in July 2003, and put into a 'parking' orbit. When it fired its solid-propellant onboard motor on 15 August, it exploded. All contact was lost, but several fragments of it were later imaged, one of which seems to be following the planned orbit. Search for Contour was officially abandoned on 20 December.

A comet is a wraithlike object: I once described a typical comet as 'the nearest approach to nothing that can still be anything'. The only reasonably substantial part is the nucleus, made up of a mixture of ices and rubble, and even this is small. The nucleus of the famous Comet Hale–Bopp, which shone so gloriously in the sky during 1997, had a nucleus 40 km (25 miles) across, and ranked as a colossus by cometary standards; had it come close to us it would have cast shadows, but unfortunately it never came anywhere near. The bright 1996 comet, Hyakutake, approached us to a minimum distance of no more than ten million miles – but it was very small, with a 1-mile (less than a kilometre) nucleus, though it has a very long tail. It was, moreover, a lovely comet, characterized by its strong green colour. If you missed it, look out for it at its next return – in around 14,000 years' time!

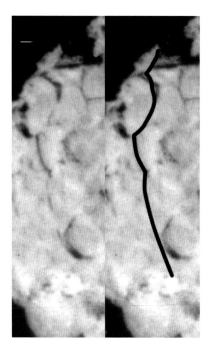

▲ *Comet Wild 2,* imaged by the Stardust spacecraft. The black line marks the location of a 2-km-long (1.2-mile) series of aligned cliffs on the comet's surface.

Quite a number of probes to comets have now been dispatched; we well remember the armada sent to Halley's Comet in 1986 – two Russian, two Japanese and one European; the ESA probe, Giotto, went right into the comet's head, and sent back close-range pictures of the nucleus. At the moment new missions are being planned, and one of these, Stardust, is well under way. It is a NASA venture, and was successfully launched on 7th February 1999. It was scheduled to collect interstellar dust, by-pass an asteroid (Annefrank), encounter a comet (Wild 2) in January 2003, and return dust samples to Earth in January 2006. The tiny, cratered Annefrank was duly imaged, and on 1 December 2003 Stardust obtained its first picture of Comet Wild – from a range of 25,000,000 km (15,500,000 miles).

It is a complicated procedure, and much can go wrong, but if all proceeds according to plan the spacecraft will approach Earth, the capsule containing the dust will be separated, parachuting down through the atmosphere and landing in Utah on 15 January 2006. Analysis of the particles will take years.

Another comet probe, Rosetta, got off to a bad start. It is an attempt to accompany a comet for months, and also to put a lander down on to the surface. The original target was Wirtanen's Comet, which has a period of $5\frac{1}{2}$ years and a moderately eccentric orbit, but construction delays meant that the launch window was missed; it had to be put off until 2004, and Wirtanen's Comet was no longer suitably placed. Instead a rather larger comet, Churyumov–Gerasimenko, was substituted; the period is $6\frac{1}{2}$ years, and Rosetta's journey there will take 10 years – an asteroid will be surveyed en route. The programme includes analysis of the materials in the cometary nucleus and the way in which the comet develops as it draws inward towards perihelion. Landing poses new problems, because the gravitational pull of the comet is negligible, and it will be only too easy for the landing vehicle to bounce off. The remedy is to use what is to all intents and purposes a harpoon capable of holding the probe down. It ought to be possible to obtain core samples, by using drills, and an on-board camera will take pictures of the surface at both visible and infrared wavelengths.

Even more dramatic is the Deep Impact mission, aimed at a small periodical comet, Tempel 1 (period $5\frac{1}{2}$ years). The original schedule opted for a launch in December 2004, and rendezvous with the comet in July 2005; the fly-by part of the mission would release a lander well ahead of time, and would photograph the impact as the lander hit the comet's surface. It was expected that the crater produced would be at least 90 metres (300 feet) wide and 23 metres (75 feet) deep, ejecting material from the nucleus and vapourizing the impactor and much of the ejecta. The distance from Earth would be 140,000,000 km (87,000,000 miles). In fact, the impact occurred exactly on schedule; the copper impactor, about the size of a domestic washing machine, crashed into the comet, while the orbiting section aimed itself to produce close-range images. A bright flare was seen, and a great deal of powdery dust was ejected. As expected, the comet was unharmed, apart from acquiring an extra crater.

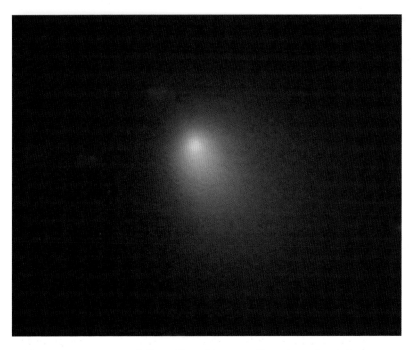

▲ **Comet Tempel 1,** imaged by the Kitt Peak 2.1-metre telescope. The red, blue and green bars in the background are stars that moved between images.

In any case, a comet such as Tempel 1 will have a limited life-span. Halley's Comet, which is much larger, has probably made about 3000 returns to perihelion, and will last for another 300 before fading away, but this is a very brief period by astronomical standards.

It used to be thought that short-period comets came from the Kuiper Belt, beyond the orbit of Neptune, while long-period comets originated in the Oort Cloud, more than a light-year from the Sun. In fact, matters may not be quite as straightforward as this, and it may be that all comets formed in the Kuiper region, in which case the long-period comets were subsequently 'pushed out' by the gravitational pulls of Uranus and (particularly) Neptune. The Oort Cloud could then be classed as a kind of comet storage region.

When will the next brilliant comet appear? We have to admit that we do not know. Comet Swift–Tuttle will be spectacular in 2126, when it will skim past the Earth – there is even a possibility, albeit a slim one, of a collision, and Halley's Comet will be at its very best at the return of 2136. But this is rather a long time to wait. Magnificent comets were rare in the 20th century; let us hope that the 21st century will be more spectacular.

25 • THE MUSIC OF
THE SPHERES

Is there any connection between astronomy and music? The answer, rather surprisingly, is 'yes', quite apart from the fact that some great astronomers and physicians have also been expert musicians – William Herschel and Albert Einstein, in particular.

It began with the Greeks. For example, there are seven notes in the musical scale (A to G) and there were seven known bodies of what we now call the Solar System: the five naked-eye planets: Mercury, Venus, Mars, Jupiter and Saturn plus the Sun and the Moon. Could this be coincidence? The Greeks – at least the earlier philosophers – believed that around the Earth, the centre of the universe, lay a series of crystalline spheres, each of which carried a planet; they were perfectly transparent, so that we could see through them, and beyond the outermost (the sphere of Saturn) lay yet another sphere, black this time, carrying the stars. The spheres revolved, each making one circuit per day, and each corresponded to one note of the musical octave, which led on to the idea of the 'music of the spheres'.

Concepts of this sort did not die at the end of the era of the Greek philosophers, and it was some time before the invisible crystalline spheres were swept away – Johannes Kepler, whose laws of planetary motion were drawn up in the first two decades of the 17th century, was among the first to abandon them, though admittedly he replaced them with an even weirder system involving 'regular solids', described in his book *Harmonici Mundi* (1619). Each planet had its own note, from the shrillness of scurrying Mercury to the deep, lugubrious tone of slow-moving Saturn.

Isaac Newton's immortal book the *Principia,* published in 1687, ushered in the modern era of science, even thought Newton himself was just as much of a mystic as Kepler or, for that matter, the ancient Greeks. (His work on alchemy and astrology occupied most of his time, and is of no value whatsoever, though as a mathematician he has never been surpassed, and probably equalled only by Einstein.)

Friedrich Wilhelm Herschel – always remembered by his Anglicized name of William Herschel – was perhaps the greatest of all observers, certainly before the age of great telescopes. Yet he was a musician who never had any real training in science, and ranked officially as an amateur. He was born in Hanover in 1738, his father was a musician in the Hanoverian Army, and young William entered the army as a bandsman. He did not care for army life (particularly after having spent one night in a ditch!) and came to England, where he remained for the rest of his life. It is not true that he deserted, as has been claimed; he was free to leave because he had been too young to be formally enlisted.

Herschel was an excellent organist, and in England he earned his living, first by copying musical scores and then by playing; remember that at that time England and Hanover were united under the same king, George II during Herschel's boyhood and then George III. In 1766 Herschel moved to Bath, then the most fashionable spa in England, and became organist at the famous Octagon Chapel. He was an instant and undoubted success, both as a performer and as a music

▼ *The Music of the Spheres.* Johannes Kepler's picture of the regular solids, linked with the orbits of the planets and with musical notes. In many ways Kepler links medieval mysticism with modern-type astronomy.

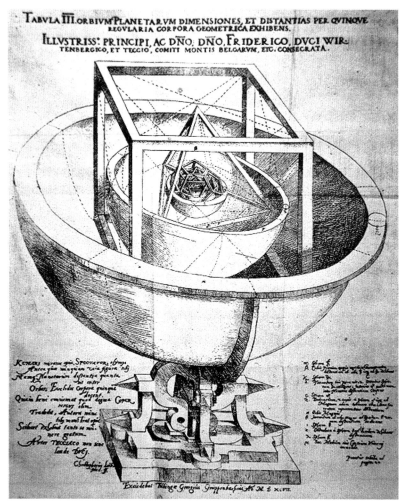

▲ *Music on* The Sky at Night. *Donald Francke (singer) and Neil Crossland (piano) kindly performed.*

teacher. He brought his sister Caroline over from Hanover; she was a singer, and intended to carve out a career, though in the event she devoted herself to her brother.

In the early 1770s Herschel became interested in astronomy and gradually this took over. He made his own telescopes and with one of these, in March 1781, he discovered the planet we now call Uranus. It was this discovery that changed his whole life, and made it possible for him to give up music as a profession; he was appointed King's Astronomer to George III, and moved from Bath to Slough, where he remained until his death in 1822. His contributions to astronomy were unequalled; for example, he was the first to give a reasonable picture of the shape of the Galaxy, and he discovered thousands of double stars, clusters and nebulae. All this has been documented on countless occasions. But what was he like as a musician?

As an organist, he was really very good indeed (what a pity we have no record of him!). As a composer he was prolific, and thought he cannot be classed in Division One he was quite high in Division Two. Recordings of his compositions are available and make for 'easy listening'.

I do have connections with the Herschel story. When William discovered Uranus he had just moved to 19 New King Street, Bath, where he remained until the move following the discovery of Uranus. His last home – Observatory House, Slough – was demolished in 1960, leaving No. 19 as the only Herschel house remaining. In 1978

I had a telephone call from a Bath resident, Philippa Savery, telling me that No. 19 was in danger of being pulled down. Philippa was not an astronomer, but she had a keen interest in all historical matters, and felt that if possible something ought to be done. I drove to Bath, met Philippa and went with her to No. 19, which was in a sorry state. We took a gamble, hired the Octagon Room, contacted the Press and the BBC, and announced the foundation of the William Herschel Society. Against all odds, the gamble paid off. One thing led to another, and today No. 19 is a well-equipped, well-run little museum. If you ever go to Bath, I recommend a visit.

My other connection was different. We put on a *Sky at Night* television programme about Uranus, and, bravely, I obtained a piano score of a Herschel composition and played it 'on the air' – live. I doubt if I did it justice, but it was great fun.

The other great scientist-musician was of course Albert Einstein, whose violin playing was well up to professional standard. I met him only once, in 1940, when I was based in Canada learning how to fly. I had a week's leave, went down to the States and was invited to a small scientific meeting at which Einstein was present. At the post-meeting reception he was asked to demonstrate his skills as a violinist – he had been taking part in an earlier concert, and had a violin with him. I accompanied him on the piano, in Saint-Saëns' 'Swan' from *Carnival of the Animals*. How I wish I had a tape of it... but there were no tapes in 1940.

▲ **Gustav Holst,** the composer probably best remembered for The Planets – *which were admittedly more concerned with astrology than astronomy!*

The most astronomically related composition is of course Holst's *The Planets*, of 1916. Actually it was more astrological than astronomical, but this makes no difference. Remember too the song *Betelgeuse*, which Donald Francke sang in our programme for January 2001; that programme ended with Catherine Galloway singing – 'It may be in my stars', from my own opera, *Perseus*.

Last – but by no means least – Brian May. He is without much doubt the best guitarist in the world today, and arguably the best there has ever been. But how many people know that he is a well-qualified astronomer who has carried out valuable research? He is characteristically reticent about his honorary doctorate. Inevitably, however, music and *Queen* take up most of his time – though when he came with us to northern Scotland for the annular eclipse of the Sun in 2003 his pictures, taken with infrared equipment, matched those of any of the professionals. As an astronomer, I am sure that we will hear more of Dr Brian May.

We discount the 'music of the spheres', but many people are familiar with 'radio noise', such as the hisses and crackles due to the Sun. Of course, the noise is produced in the receiver of the radio telescope, but the Sun is responsible, and we have similar 'noise' from bodies far out in the universe. There has even been a plan to record the noise of the winds in the thin atmosphere of Mars. This is not music, and in any case the probe carrying the recorder failed, but one day there will no doubt be concerts on Mars!

26 • ROVERS TO MARS

So far as Mars was concerned, the year 2004 began with mixed fortunes. It started with the loss of Beagle 2, the British probe master-minded by Professor Colin Pillinger and his team. Until the last moment all seemed well. Beagle was carried by the Mars Express orbiter; Beagle was separated at the right time – I was with the team at the Rutherford Appleton Laboratory when this happened, and there was loud cheering. Beagle went on its way in free fall, and presumably landed on Mars on Christmas Day. That was that. Nothing more was ever heard from it, total silence, and finally, after weeks of fruitless hunting, Beagle was given up as lost.

What happened? Possibly the parachute failed, and there were no retro-rockets to slow the probe down before landing. It may be that Beagle landed in an area or at an angle which stopped it from transmitting. We may never know – unless the probe, or bits of it, are found either by photograph or by future explorers. As I said during a broadcast, for a successful Mars mission you need both skill and luck. Colin Pillinger had the skill, but on this occasion he didn't have the luck.

Against this, Mars Express gave no trouble at all, and soon began sending back amazing pictures as well as a stream of information. Of course one of the main requirements is to search for water, and Express's first triumph was to detect the spectral sign of water in the south polar cap. We were already sure that water ice existed, but the only reliable guide had been the detection of hydrogen, so that the Express findings were much more positive. Water vapour was detected in the Martian atmosphere, which was another new if not

◀ **Beagle 2.** A model of the ill-fated British Beagle 2 probe, photographed by Chris Lintott.

▶ **Discussing the rovers to Mars.** In my study (the studio) are Garry Hunt and Andrew Coates. The Commanding Officer sign indicates the usual sleeping place of my beloved cat Jeannie!

unexpected development. Then, down came the two NASA landers – Spirit in the large crater Gusev, and Opportunity in Meridiani Planum on the opposite side of Mars. Both landings were completely successful. Both rovers had problems after arrival, however, but these were soon put right, and the two vehicles began moving around, making measurements and carrying out analyses of the rocks. They were programmed to remain active for 90 days, the maximum period during which they could take in enough solar power. They were much more elaborate than Sojourner, the only previous rover, and they could travel further. By February 2004 there were four spacecraft in or near the Red Planet: Mars Global Surveyor and Mars Express in orbit, and Spirit and Opportunity on the surface. Express even managed to photograph Spirit, though there was still no sign of Beagle 2.

The landing sites had been carefully chosen. Crater Gusev may once have been a lake; several channels debouch into it, and in the remote past these channels can hardly have been anything other than torrents of water. Spirit found that the red surface layer was 'dust' – oxides, as fine as talcum powder; there is no sand on Mars because of the absence of quartz from which sand is produced. Spirit made its way to a large rock, nicknamed Adirondack, and used the Rock Abrasion Tool to grind off a portion of the surface, several centimetres (about $1\frac{1}{2}$ inches) across. Dr Squyres, one of the main investigators, commented: 'What you are seeing there is a beautifully cut,

▲ **Opportunity on Mars.** *This image was taken by Opportunity's panoramic camera. The rocks, which line the inner edge of a small crater, are only about 10 cm (4 inches) tall.*

◄ **Spirit on Mars.** *The first rock to be examined by the Spirit rover was Adirondack. It was chosen for its smooth, easily grindable surface.*

almost polished rock surface. It looks very much like a volcanic rock, a basalt…we find compelling evidence that, in fact, what we are looking at is a volcanic basaltic rock. So the RAT has revealed the interior of this rock. We know what it is – a piece of volcanic stuff. And it is time to move on.' Up to that time Spirit had covered almost 20 metres (70 feet), easily breaking Sojourner's record. The next target was the crater Bonneville.

Opportunity came down in a small impact crater in Meridiani Planum, and soon began a photographic survey of the outcrop of the crater. The 30-metre (90-foot) long bedrock formation proved to have fine structure, with layers only a few millimetres (a few tenths of an inch) thick and with embedded spherical grains whose origin is still unclear. They might be impact or volcanic ejecta formed in situ when liquid percolated through the rock. Meridiani Planum had

been selected because of the presence of grey hematite, which is a mineral often deposited by iron-rich water.

We know that liquid water cannot exist on Mars today, because the air pressure is too low, but there were seas once, as well as violently active volcanoes; moreover Mars had an overall magnetic field, which has to all intents and purposes disappeared. But can there be liquid water below the surface? Quite probably. Mars Express carries radar which can probe down to a depth of three miles, so that before long we may have the answer to one of the most important problems. In fact, we may know by the time that you read this page!

We are not sure whether any volcanism lingers on. Major eruptions belong to the past, but there may well be fumaroles – and we cannot be sure that the volcanoes are extinct; Mars goes through very marked changes in climate and in its general condition, and one day the volcanoes may again erupt.

Life – well, we can rule out little green men and Percival Lowell's canals, but it would be premature to dismiss Mars as completely sterile. We may know when we send up a 'sample and return' probe, which is scheduled for around the year 2011. Of course there will be missions before then: NASA has plans, and the European Space Agency has its 'Aurora' programme. This is not to suggest anything in the nature of a space race, as in the Cold War era, but certainly there is healthy rivalry. President George W. Bush has laid down a timetable, leading on to a manned expedition as early as 2030. People with nasty, sceptical minds (such as me) suspect him of 'doing a Kennedy' in the year before the next US Presidential election, but there is always the chance that on this occasion he is sincere. Time will tell.

In any case, the investigation of Mars will continue unabated. There is so much to learn, and the Red Planet, so much more like Earth than any other planet in the Solar System, presents us with a challenge we will not ignore.

27 · HUNTING FOR WIMPS

On 25 March 2004, I introduced *The Sky at Night* programme from a most unusual setting. Instead of being in a BBC studio, or in my study at Selsey, I was in a miner's cage, on my way down to the bottom of a salt-mine at Boulby, near Whitby in Yorkshire. There I would find an elaborate scientific laboratory, manned by scientists who were searching for WIMPS – strange particles coming from the depths of space.

It was not the first time that the programme had 'gone underground'. Many years earlier we had been down Homestake Mine, in South Dakota, where efforts were being made to track down neutrinos sent out by the Sun. Neutrinos are elusive, to put it mildly; they have no electric charge, and until recently it was thought that they lacked mass. The best way to detect them is by interactions with ordinary matter, and this is what was being done at Homestake. It was essential to go underground, to get away from cosmic rays – which are not rays at all, but high-velocity atomic nuclei. Cosmic rays cannot penetrate a thick layer of solid earth, but neutrinos can. The Homestake experiment is still going on, and has been very successful even if the results were not quite what had been expected. But there were two important differences between Homestake and Boulby. Homestake was a gold-mine, Boulby a mine for salt and potash. And the Boulby workers were not particularly concerned with neutrinos; WIMPS were much more in their minds. However, it was still essential to go deep down, and as well as making sure that cosmic rays did not interfere, the equipment also had to be shielded against natural radioactivity in the Earth's rocks.

But what are WIMPS – Weakly Interacting Massive Particles – and if we cannot see them, how do we know that they exist? This brings me on to the whole question of 'dark matter' in the universe, possibly the most significant problem facing astronomers today.

The pioneer in the field was Fritz Zwicky, a Swiss citizen who spent most of his working life in America. He was a curious character, to put it mildly. Anyone who questioned his theories was automatically classed as a mortal enemy, and he described his contemporaries as 'spherical bastards' because he maintained that they were bastards whichever way you looked at them. He ordered his night assistant to fire bullets through the opened observatory dome, because he thought that this might improve the quality of seeing (it didn't). But of his brilliance there was no doubt at all. He was the first to undertake a systematic search for supernovae in external galaxies, with great success. He studied clusters of galaxies, and worked out from their movements that the clusters should be unstable; something or other was acting as cosmic glue, preventing the galaxies from flying apart. Later there came revelations about the ways in which galaxies rotate. Spiral systems rotate round their centres; our own Galaxy is no exception – the centre lies around 26,000 light-years from us, and the Sun takes some 225,000,000 years to complete one circuit (a period sometimes called the 'cosmic year'). It is natural to suppose that the systems close to the centre move more rapidly than those further out. This is what happens in the Solar System, as was first realized by Johannes Kepler four centuries ago; thus Mercury moves round the Sun at a greater speed than Venus, Venus faster than the Earth, and so on. But in galaxies Kepler's Laws do not apply, and the only explanation is that the main mass of the galaxy is not concentrated in its core; instead, it is spread around.

All in all, most astronomers now believe that the material we can see – stars, planets, galaxies – makes up only a tiny fraction of the total matter in the universe. The rest is invisible. Dark matter accounts for perhaps 25 per cent of the total; the rest is attributed to the even more mysterious 'dark energy'. But let us leave dark energy for the moment, and see what we can learn about dark matter. Is it something which we might be able to observe, or is it entirely different from anything in our experience?

▼ **In Boulby Mine.** To the far left, the cameraman Dave. I am flanked by Bernard Carr and Carlos Frenk; beyond Carlos, Jane Fletcher, Mary and the Boulby nurse. Chris Lintott is standing behind me.

◄ **Inside Boulby Mine.** *A photograph of one of the tunnels making up the vast 1400-km (900-mile) long underground network.*

▼ **The exterior of Boulby Mine,** *a salt-mine near Whitby, Yorkshire.*

We have to account for the 'missing mass', and all sorts of candidates have been proposed. Neutrinos were favoured at one time. They are so plentiful that if a neutrino had appreciable mass we would presumably have the answer to our problem. Alas, the results at Homestake Mine and elsewhere show that although neutrinos do not have zero mass, what they do have is quite unable to explain the missing mass of the universe. Vast numbers of low-mass stars are equally inadequate, and locking up matter inside black holes does not seem feasible. WIMPs are much more promising – if only we could determine their nature.

The trick is to persuade them to interact with ordinary matter, which is what the Boulby scientists are trying to do. Together with three leading WIMP-hunters – Dr Nigel Smith (Rutherford-Appleton Laboratory), Professor Carlos Frenk (Durham University) and Professor Bernard Carr (Queen Mary College, London) – I stepped out of the cage as we reached the bottom of the shaft, and looked around with immense interest. We were a mile down; Boulby is the deepest mine in Britain, and it is in full operation. There are 1400 km (900 miles) of tunnels, extending far out below the sea. The dark matter researchers were cordially received, and have been given every possible help. Their main laboratory is indeed well equipped (apart from the absence of a toilet…).

One detector, known as ZEPLIN, uses a tank filled with four kilograms of liquid xenon. (Xenon in its gaseous form is a trace constituent of the Earth's air, and is decidedly reluctant to interact with atoms of other kinds.) If a WIMP collides with a xenon nucleus in the tank, the energy from the recoil of the nucleus produces a spark of light. This light is detected by photomultiplier tubes, which amplify the signal. The problem is that in the tank only about one hit per day is expected, so that the experiment is amazingly delicate. Obviously there is more to it than I can describe here, but the underlying principle is sound, and more sophisticated ZEPLINs are being planned. Another detector is NAIAD (Nai Advanced Array Detector), which uses an array of sodium iodide crystals (hence Nai; Na is the chemical symbol for sodium, I for iodine). And DRIFT (Directorial Recoil Identification from Tracks) will attempt to find out the directions from which WIMPs come.

This is all very well, but up to now the results have not been unambiguous, and we are not a great deal further forward. At least a start has been made, however, and there are other research bases working along the same lines, mainly in America, Italy and Japan. If we can prove that WIMPs really do interact with ordinary matter, albeit rarely, we will know that we are on the right track. Then will be the time to start trying to understand just what they are.

'Dark Energy' presents us with another set of problems. We know that every group of galaxies is receding from every other group, so that the entire universe is expanding; the greater the distance, the greater the velocity of recession. We are now fairly confident that the expansion started after the Big Bang, 13.7 thousand million years ago, when the entire universe came into existence. It was logical to assume that the rate of expansion slowed down because of the force of gravity, but it now seems that at a sufficiently great distance the rate of expansion is increasing, so that in effect we have an 'accelerating universe'. (This has been established by observations of supernovae, which reach about the same maximum luminosity, but when sufficiently remote appear fainter than they would if the rate of expansion had been constant from the very beginning – so that they are more distant than would have been expected.) Einstein had introduced a term acting in opposition to gravity. He called this force of repulsion the 'cosmical constant' but later abandoned it, and even said that it had been the greatest mistake of his career. It is now starting to look as though he had been right originally, and that dark energy really does correspond to his cosmological constant. Unfortunately, we have to admit that as yet we do not have the faintest idea of what dark energy may be, and there are some cosmologists who are openly sceptical about it, even though the accelerating universe is hard to explain without it.

I emerged from Boulby mine full of admiration for the dedicated scientists at work there. Some day – perhaps soon – they may solve the WIMP problem. Meanwhile, it is sobering to reflect that despite all the beauty and brilliance of the stars and galaxies, perhaps 99 per cent of all matter in the universe is concealed from our inquiring eyes.

28 · STAR-BIRTH

Look up into a clear night sky, and you will see stars of many kinds. All are suns, and there is a tremendous range in both luminosity and size. We know stars which are millions of times more powerful than our Sun, while others are feeble; represent the Sun by a pocket torch, and we will find both searchlights and glow-worms. Some are big enough to hold the entire orbit of the Earth round the Sun, while others are no more than a few miles across.

▼ **Star-forming region:** the 'Pillars of Creation' in the Eagle Nebula (M16 in Serpens), photographed with the Antu section of the Very Large Telescope in Chile.

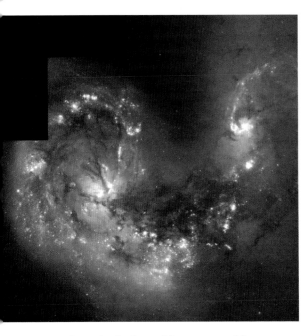

◀ *The Antennae galaxies (NGC 4038/4039),* imaged by the Hubble Space Telescope. The bright blue regions are areas of great star-birth activity.

But why is there this range, and how are stars born? Also, how old are they?

Obviously we must start with the Sun, which is the only star close enough to be studied in detail. It is around 1,390,000 km (864,000 miles) in diameter, and it sends us virtually all our light and heat; the surface temperature is almost 6000°C, and near the core the temperature rises to an almost unbelievable value of 15,000,000°C. The Sun is made up of gas, with hydrogen the most plentiful element. It is not 'burning' in the conventional sense; a Sun made up of coal, and radiating as fiercely as the Sun actually is, would turn to ashes in a few million years, and the Sun is much older than that. We are confident that the age of the Earth is about 4.7 thousand million years, and that our world (and the other planets) was formed from a cloud of material surrounding the youthful Sun; this puts the Sun's age back to something like 5000 million years. It shines by nuclear reactions. Deep inside the globe, nuclei of hydrogen are combining to form nuclei of helium. It takes four nuclei of hydrogen to make one nucleus of helium; every time this happens, a little energy is set free and a little mass is lost. It is this energy which makes the Sun radiate, and the mass-loss amounts to 4,000,000 tons per second. This may sound a great deal, but it is not much compared with the Sun's total mass. There are certainly short-term fluctuations in output, but the overall picture has not changed significantly since the first advanced life-forms flourished on the Earth. Changes will come eventually, but our Earth will remain habitable for at least a thousand million years yet, unless we ourselves indulge in wars which will make the whole globe radioactive and barren.

The stars are light-years apart, but the space between them is not empty; there is always a certain amount of material, though the density of this interstellar material is lower than the best vacuum we can produce in our laboratories. However, we do see patches of material, which we call nebulae, and these are the birthplaces of stars. Some nebulae are visible with the naked eye; probably the most famous is the Great Nebula in Orion, in the Hunter's Sword, just south of the three bright stars making up the belt. It is 1500 light-years away, and is lit up by the stars of the so-called 'Trapezium'. We cannot see through to the centre of the nebula, but we know that powerful stars exist there, detectable by their infrared radiation, which passes unhindered through the gas and dust.

A nebula is not static; there is always turbulence, due mainly to the effects of nearby stars. This means that clumps of material can form. As soon as these denser areas have accumulated enough mass, extra material can be drawn in by the force of gravity. The embryo star shrinks, and its interior heats up. If there is enough mass, the core temperature rises ten million degrees, and this is enough to trigger nuclear reactions, mainly involving the conversion of hydrogen into helium. A true star has been born.

Everything depends upon the initial mass. If the core temperature never rises sufficiently for nuclear reactions to begin, the star merely shines feebly for a very long time before becoming cold and dead; it may be said that a 'brown dwarf' is a star that has failed its entrance examination (the name is rather misleading, because when seen clearly the colour would be red, not brown). If the mass is great enough, as with the Sun, the star is at first irregularly variable, and is surrounded by a cocoon of material. At last the cocoon is blown away, and the star settles down to a long period of stability – this is what we call the Main Sequence stage. When the store of available hydrogen 'fuel' is used up, the star has to change. The core shrinks and becomes even hotter, different nuclear reactions begin, and the star's outer layers expand. It has left the Main Sequence, and has become a giant. When the Sun turns into a giant, the inner planets will be destroyed; even if the Earth escapes total destruction, it will be left a scorched cinder from which every trace of life has vanished.

A star with much greater initial mass will be born in the same way, but everything will happen much more quickly. The mass-loss is much greater than with the Sun, and the life-span is much shorter. When the hydrogen fuel is used up, the star 'implodes', and then explodes as a supernova, sending most of its material into space and leaving only a tiny, super-dense remnant. The last supernova observed in our Galaxy was that of 1604, though in 1987 one blazed out in the neighbouring galaxy known as the Large Cloud of Magellan – about 170,000 light-years away.

The material blown away by supernovae is a mixture of elements, and it is from these elements that new stars are produced, but always the essential process is the same: a nebula is a stellar nursery, and inside some nebulae it seems that star formation is going on at a furious rate, though of course the birth of a new star is bound to be a leisurely process by conventional standards. We can see wildly active 'starburst galaxies'; we can see the results of collisions

▼ **The Orion Nebula,** a star-birth region. It is illuminated by light from the hot 'Trapezium' stars, visible at top centre of this Very Large Telescope image.

between galaxies – such as the famous Antennae, where the distortions in shape are so vivid. In the far future there will be a collision between our Galaxy and the Andromeda Spiral, which is still over 2 million light-years away, but is inexorably approaching. However, the collision will not happen until our Sun has ceased to exist in its present form. In a collision, the individual stars seldom meet head-on, but the tenuous material between them is colliding all the time, and increased star formation is the result.

We now have magnificent pictures of these clouds of gas and dust; they are much less placid than they look. It is fascinating to realize that some 5 thousand million years ago our Sun was born inside one of these nebulae.

29 • TRANSIT OF VENUS

I have many fond memories of *The Sky at Night* programme. One of the happiest of all was that of 8 June 2004, transmitting live from my observatory at Selsey – or, rather, the garden round the Observatory. Venus was due to pass in transit across the face of the Sun, and to my surprise as well as relief the BBC agreed to send down a team to cover it. After all, nobody living had ever seen a transit of Venus. The last dated back to 1882.

Venus takes 225 days to complete one orbit round the Sun, at a mean distance of 108,000,000 km (67,000,000 miles). When it is almost between the Sun and the Earth it is of course 'new' with the dark hemisphere facing us, and cannot be seen – unless the alignment is exact, when Venus will appear as a black disk passing across the brilliant solar photosphere. On most occasions Venus will be 'new' either above or below the Sun in the sky, and will avoid transit. Mercury transits more often, and did so in May 2003, but appears as nothing more than a black speck, because it is not only much smaller than Venus but also much further away. We had observed the Mercury transit; we hoped for equal success with Venus.

A transit of Venus is a leisurely affair. On 8 June it would begin soon after sunrise, and would not end until midday, so there was no

▼ **Observing the transit,** *using a special solar filter* with telescopes large and *to watch Venus' progress* small. At right a guest is *across the Sun's disk.*

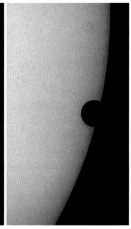

▲ **Venus at ingress,** *a small black notch. Jamie*
moving on to the Sun's *Cooper took this series of*
disk. It appears initially as *images from Selsey.*

frantic haste as there is with a total eclipse of the Sun. We were anxious to follow the whole phenomenon, which meant not only a clear, cloudless sky but also a horizon unobstructed by trees or buildings. Earlier in the week I went out at sunrise, with Chris Lintott, and checked. All was well. The Sun rose conveniently over a clear area, and as it gained altitude it soared well above all the other low trees. Also, the weather forecast was good.

I had sent out a goodly number of invitations; join us for the transit, have some refreshments around breakfast-time, and after the end of the transit retire to Farthings, my home, to enjoy glasses of wine. It would take time to set up the equipment, but there was plenty of space in the ground beside the Observatory.

First contact was due at 06.14 British Summer Time (05.14 GMT), and Venus would be fully on to the solar disk by 06.33. Everyone was anxious to follow this, because of the infamous Black Drop effect. As Venus moves on to the disk it seems to draw a strip of blackness after it, and when this strip disappears the transit has really begun – making it impossible to make an accurate timing of ingress; this effect was what had ruined Halley's method of using transits to measure the distance between the Earth and the Sun. It had always been thought that the Black Drop was due to Venus' atmosphere; this is not so – but we all wanted to see what would happen.

By six o'clock the garden was bristling with telescopes and other instruments. Some were elaborate; for example Brian May was planning to observe at infrared wavelengths, and the various photographers were in their element. The BBC team had arrived, and were recording interviews with some of the observers. I think that all together there were about a hundred people there – all waiting.

Conditions could not have been better – a perfectly transparent sky, a pleasant temperature, and no wind to shake the equipment. I was not proposing to attempt anything really scientific, my role was to comment, and I projected the Sun using the faithful 3-inch refractor that I had bought for £7.10s when I was a boy (in 1934!). We waited. At the appointed time a tiny notch appeared. Slowly it became more obvious – and then came the really significant moment. Yes, there was a Black Drop. It was impossible to tell just when the disk of Venus was fully silhouetted; I suppose that the effect lasted for between one and two minutes, and then it disappeared. The transit was underway.

Two things struck me. First, the size; of course it had been worked out and I knew what to expect, but Venus was so different from the tiny speck of transiting Mercury. Secondly, the utter blackness of Venus. Unfortunately there were no really large sunspots to act as comparisons, but Venus looked like a disk of ink.

For the next few hours everyone was busy, and everyone seemed to be getting good results. The sky remained clear – not a cloud in sight, and no trace of sea mist. Near 'maximum transit' many people used filters to look at Venus with the naked eye, and certainly it was easy to detect. I wondered what observers elsewhere were

▲ **Projecting the Sun;** *Venus is seen as a small black disk. Note how tiny* *the planet appears against* *the Sun (though not as tiny* *as Mercury at transit).*

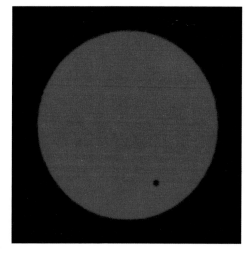

◄ ▲ *The transit of Venus,* as imaged in visible light by John Fletcher. The photograph at left was taken through a red filter.

doing. There was a gathering at the South Downs Planetarium in Chichester, and there were others all over the country. Not everyone was lucky. John Fletcher had decided to watch ingress with the equipment at his observatory at Tuffley, near Gloucester – and was clouded out. He promptly got into his car and set out for Selsey, arriving in time for the best view. (I hope he encountered no speed cameras on the way.)

Around a quarter past nine Venus reached its maximum distance from the centre of the Sun, and began to crawl slowly towards the limb. By now there was a sort of carnival atmosphere in the garden; all the photographs and experiments were going well – and of course all of us knew each other, though some passers-by joined in, and there was a visit from the schoolboys of Prebendal School in Chichester – who were vastly intrigued. Time seemed to pass quickly. The leading edge of Venus was due to touch the Sun's limb at 12.07 BST, and the following edge at 12.26. Would we see the Black Drop at egress?

Cameras and telescopes poised, we waited. Final egress – and yes, there was the Black Drop. It lasted for perhaps half a minute, and was gone. Venus was passing off the solar disk, soon it was a tiny notch – and at 12.23 I lost it.

I think we were all sorry when the transit was over; it had indeed been a memorable experience. I will not pretend that any valuable scientific work had been done – transits of Venus have lost their importance – but this did not make it any the less enjoyable, and as we packed up our gear and made our way into the house for wine and 'nibbles' it was clear that everyone was well satisfied.

There are occasions when everything goes right. This was one, and Nature had been in its kindest mood.

30 • ENCOUNTER
WITH PHOEBE

Saturn is, in my view at least, the loveliest object in the sky. Its glorious, icy rings are in a class of their own. But do not forget Saturn's family of satellites. Over 30 are now known, but most are very small. Before the Space Age, there were nine known satellites:

Name	Discoverer	Mean distance from Saturn thousands of km (thousands of miles)	Orbital period days hours mins			Diameter km (miles)
Mimas	Herschel, 1789	186 (115)	0	22	37	420 (261)
Enceladus	Herschel, 1789	238 (148)	1	8	53	512 (318)
Tethys	Cassini, 1686	294 (183)	1	21	18	1046 (650)
Dione	Cassini, 1686	378 (235)	2	17	41	1120 (696)
Rhea	Cassini, 1672	528 (328)	4	12	25	1528 (950)
Titan	Huygens, 1655	1223 (760)	15	22	41	5151 (3201)
Hyperion	Bond, 1848	1480 (920)	21	6	38	360 (224)
Iapetus	Cassini, 1671	3540 (2200)	79	5	56	1435 (892)
Phoebe	Pickering, 1898	12,955 (8050)	550	10	50	227 (141)

Titan, of planetary size, was known to have a dense atmosphere. The remaining satellites were assumed to be icy and cratered, and this was confirmed when the Voyager probes passed through the Saturnian system in 1980 and 1981. Iapetus was of special interest; its leading side was dark, and the trailing hemisphere was conventionally bright and icy. Was this due to a deposit collected from space, or was the darkness caused by material welling up from below the surface? Nobody knew (in fact, we are not sure even today).

The outermost satellite, Phoebe, was different. It was a long way out, and moved in a wrong-way or retrograde direction; also, it was thought to have an axial rotation period of only 9 hours. The Voyagers went nowhere near it, and only one image, from Voyager 2 on 4 September 1981, showed a feature that might well be a crater. It was tacitly assumed that Phoebe was an ex-asteroid, captured from the inner belt – and that it would turn out to be icy.

The Cassini probe was launched from Canaveral on 15 October 1997. (The name honoured the Italian astronomer, who discovered four of the satellites. It carried a smaller probe, Huygens, scheduled to make a controlled landing on Titan.) Following swing-by encounters with Venus (21 April 1998), Venus again (20 June 1999), Earth (16 August 1999) and Jupiter (30 December 2000), the spacecraft finally reached its target in June 2004. After some manoeuvres it entered an orbit round Saturn, and on 11 June by-passed Phoebe at

a range of 2071 km (1287 miles), sending back detailed images of the surface of the satellite.

The results were surprising. It was clear that Phoebe was not in the least like the asteroids that had been imaged in the main belt. It was an ice-rich globe, part of which

▲ **Phoebe,** *imaged by the Cassini spacecraft. It does not look in the least like a main belt asteroid; it seems to have been born in the outer reaches of the Solar System.*

was overlaid by dark material reflecting a mere 8 per cent of the sunlight falling upon it, but it was scarred by craters (some of which were large – over 50 km (30 miles) across), grooves, ridges and chains of pits; tiny pits were very numerous, some of them associated with bright streaks recalling a miniature lunar ray system. There were bright streaks on steep slopes, possibly where loose material had slid downhill when the surface was shaken by impacts of meteoroids. There were blocks on the surface, and areas which were particularly dark (rich in carbon compounds?). So why was Phoebe so unlike anything that had been expected – and what was its origin?

It did not fit with the idea of its being a body captured from the main asteroid zone between the orbits of Jupiter and Mars. Certainly it was not a 'regular' satellite, because of its retrograde motion. But there was another possibility. Phoebe might have originated in the Kuiper Belt.

By now we know of many hundreds of asteroid-sized bodies moving in the outer reaches of the Solar System, near and beyond the orbits of Neptune and Pluto. Some, such as Quaoar and Sedna, are

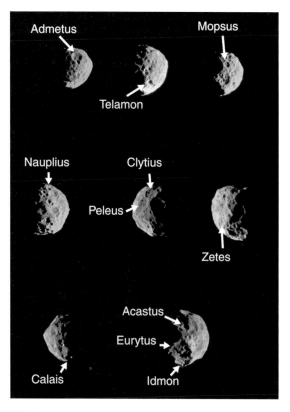

◀ **Craters on Phoebe,** *as imaged by the Cassini spacecraft in June 2004. The names are those provisionally chosen by the International Astronomical Union, after the Argonauts of Greek legend.*

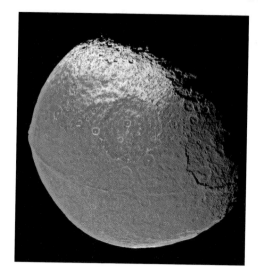

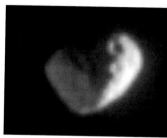

▲ **Janus,** *from Cassini.*
There appear to be two
large craters near the
day–night line.

◀ **Iapetus,** *from Cassini.*
It is heavily cratered, with
a curious equatorial ridge.

over a thousand miles in diameter – not very much smaller than Pluto, and it is now generally believed that Pluto should be ranked as a large Kuiper Belt Object (KBO) rather than a bona-fide planet. Undoubtedly objects do sometimes break free from the Kuiper Belt and invade the inner Solar System, and Phoebe might well be one of these. We do not yet have good, close-range images of the surfaces of any KBOs, but we would expect them to resemble what we have found with Phoebe.

After the fly-by, Cassini went on to thread its way between two of the rings, and then began a four-year tour of the system. Most of the main satellites will be imaged, but there will be no more close approaches to Phoebe, so that we must be content with the information we obtained on 11 June this year. At least it seems fairly definite that Phoebe is unlike any other member of Saturn's family, and the only one to be a probable escapee from the remote Kuiper Belt.

31 • THROUGH THE RINGS OF SATURN

Look at Saturn through a telescope – even a small one – and when the ring system is 'wide open', as it was during the summer of 2004, you will see that the two main rings, A and B, are separated by a gap known as the Cassini division. It is almost 5000 km (3000 miles) wide; G.D. Cassini discovered it in 1675, and until recently it was assumed to be empty. The B ring is much the brightest in the system; the A ring contains a 320-km (200-mile) gap, the Encke division, in which moves a tiny satellite, Pan. Beyond the main system come much fainter rings, F, G and E.

As we have seen, the Cassini spacecraft, justly named in honour of the Italian astronomer who discovered the main division in the rings (as well as four of Saturn's satellites – Iapetus, Rhea, Dione and Tethys), by-passed the outer satellite Phoebe on 11 June 2004, and sent back pictures showing that Phoebe was not quite the sort of world that had been expected. Cassini's next manoeuvre, on 1 July, was to fire its main engine to slow the spacecraft down, allowing it to

▼ **Saturn,** from the Hubble Space Telescope. The Encke division (near the rings' outer edge) and the wider Cassini division (nearer the centre) are clearly visible.

▶ *Prometheus and the F ring.* *The image, taken by Cassini, appears to show the shepherd satellite Prometheus pulling a strand of material away from the ring.*

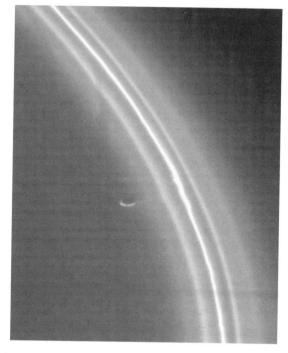

be captured by Saturn's gravity and put into a path round the planet, so that it could begin its 74 orbits – involving 44 close passes of Titan, the major member of Saturn's family of moons. All went well. The firing was faultless, and Cassini, carrying the Huygens satellite, passed between the ring system and the planet, a mere 19,980 km (12,416 miles) above the Saturnian cloud-tops. This was probably the most dangerous part of the entire mission, because Cassini had to cross the ring-plane, and nobody knew how many tiny, high-velocity particles it would encounter. Eighty-five minutes before the engine burn, Cassini rotated so that its main antenna dish was forward; this would, it was hoped, act as a shield during the crossing. Cassini had to pass twice through a known gap between the F and G rings, first while ascending just before the engine burn and again while descending after the burn.

It was a tense time for the waiting scientists. An impacting particle would release a tiny 'puff' of gas; equipment aboard Cassini could record this in the form of sound. As the ring-plane was approached, the number of impacts increased. In the plane itself, the rate reached 100,000 per minute – but the masses of the impactors were so low that no damage was caused. All the same, the scientists breathed sighs of relief when the ring-plane had been safely negotiated. 'It was like listening to a hailstorm on a tin roof,' was one comment.

▲ **Saturn's rings.** *Here the Cassini spacecraft is looking through rings A, B and C (from top to bottom). The Cassini division is in the centre.*

One interesting picture sent back during the manoeuvre showed that dusty particles were being pulled away from the F ring by the 100-km (62-mile) satellite Prometheus. It was known that Prometheus and a similar small satellite, Pandora, act as 'shepherds' to the F-ring, keeping the particles firmly in their stable orbits.

Images show that, as expected, there are thousands of ringlets and narrow gaps; the rings are made of water ice particles, but those closest to the planet are apparently contaminated with iron or carbonaceous materials. There was also a surprising amount of dark matter – 'dirt' – in the ring gaps and also in the Cassini Division. This dark matter appeared to be very like the deposits found on Phoebe. Another instrument on Cassini detected large quantities of oxygen at the edge of the rings, and the origin of this oxygen is a real puzzle.

Cassini did not neglect studies of the globe. Saturn certainly has a core at a relatively high temperature, but its precise composition is unclear. Round the core are deep 'oceans' of metallic hydrogen and molecular hydrogen, but we do not know much about this metallic layer because even in our latest laboratories we cannot match the incredibly high pressure found there. It was discovered that the winds on Saturn decrease quickly with increasing altitude above the cloud-tops: they fall off by 480 km/h (300 mph) over an altitude range of 320 km (200 miles) in the upper stratosphere. This is the first time that windspeeds have been measured at altitudes so high in the Saturnian atmosphere.

Measurements were also made of Saturn's magnetic field, which is much stronger than that of the Earth though far weaker than that of Jupiter. One unusual point is that the magnetic axis is within one degree of the axis of rotation, so that the magnetic poles and the poles of rotation coincide.

What next? Well, Cassini has a four-year programme of activity ahead of it; there will be passes of all the main satellites, and of course the climax of the whole mission with the Huygens touch-down in January 2005. But at an early stage it was clear that Cassini, the most complex spacecraft ever launched, was a resounding success.

32 • *ROBOTIC TELESCOPES*

Around 10 years ago I had a telephone call from Professor Mike Bode, head of the Astrophysics Department at the Liverpool John Moores University. I knew him, of course, and I had even spoken at the University. His call was concerned with a new venture – the idea of setting up fully automated telescopes in various parts of the world. Finance had to be raised; would I join the University committee to give some advice on fund-raising?

Of course I said 'yes', and I went to a couple of preliminary meetings. Then, to my intense surprise, I was asked to become Chairman of the Committee. Why me? I had no official link with the University; I lived in the further part of England; and I was, to put it mildly, no expert with respect to automated telescopes. But I remember saying 'I feel very honoured, and if you want me to be Chairman, so be it. But I must stress that I will be a proper Chairman, not a puppet, and I'll do things my way. Up to you.'

I was duly elected. It meant two years' hard work, because fund-raising for a project that has nothing to do with killing people, and

▼ **The Liverpool robotic telescope,** at the Observatorio del Roque de Los Muchachos, on La Palma. It is just above the clouds!

cannot be taxed, is a daunting problem. But in the end the money was raised – more than enough to get the project started – and then came the question of getting the telescopes made. In America? That was the initial suggestion, but I was totally against it; a Liverpool firm was found, able to do the job well. The result was extra employment for Liverpool, and it meant that the whole programme would be completely British. When all was settled, I called a final meeting and gave the results. The committee had done what it had set out to do; its work was done, and it disbanded. Since then I have had no official involvement – up to the time of the August 2004 *The Sky at Night* programme – but I will always be glad that I was concerned in those early days.

The first Liverpool robotic telescope came into use in April 2004, at La Palma, in the Canary Islands. Its optical design is conventional enough; it has a 2-metre mirror, and is of the Ritchey-Chrétien type. The mounting is altazimuth (as with most large modern telescopes) and, of course, it uses a CCD; it can image objects down to a magnitude of 30 in an exposure time of little over one minute. It can slew very rapidly from one part of the sky to

▶ *The Liverpool telescope at La Palma.* It was designed and built by the Liverpool firm Telescope Technologies Ltd.

◀ **Spiral galaxy NGC891,** *photographed by Nik Szymanek with the Faulkes robotic telescope – a twin of the Liverpool telescope.*

▶ **The Liverpool robotic telescope,** *La Palma. The photograph was taken by Chris Lintott in 2004, the year in which the telescope came into use.*

another, and it does not need an observer standing by it. It is purely robotic.

But why is another 2-metre (80-inch) telescope needed at all? There are many reflectors larger than that, including some on La Palma itself. The answer is quite straightforward. Telescope time is at a premium; up till now an observer who wants to carry out a definite programme on a giant telescope has to apply for it months in advance, and even if his or her application is successful he or she will probably get a limited period, which may be cloudy.* There has been an urgent need for a major telescope which can be brought into action almost at a moment's notice in order to observe unexpected phenomena. Consider, for example, the sudden appearance of a supernova. Obtaining data at once is of vital importance, and this the robotic telescope can do; it can be 'instructed' from afar, probably by an astronomer hundreds of kilometres away, and it can make all the observations needed. There are also gamma-ray bursters, probably the most violent events that can occur in the universe at present; they can leave visible signs which do not last for long, and the robotic telescope can at once image the aftermath before it disappears. An interesting comet that arrives on the scene without the slightest warning

---

*Chris Lintott found this out in August 2004. He had been given a week with the James Clerk Maxwell Telescope on Hawaii. On the first of his seven allotted nights the telescope was undergoing minor repairs, and the following six nights were completely overcast. He returned home muttering 'Dear me! Tut tut!' and 'How annoying!'. At least it meant making another trip to Hawaii!

can be imaged within minutes. Then there are NEAs or Near-Earth Asteroids, which are much more numerous than used to be thought, and which are potentially dangerous. Thanks possibly to the recent spate of 'disaster' science fiction films, such as *Deep Impact*, serious efforts are now being made to identify NEAs and monitor them; if we knew of an impending collision with a midget asteroid we might be able to do something about it, by using a nuclear device to deflect it away from Earth, assuming that we had sufficient advance warning. This principle was tested almost as soon as the Canary Islands telescope came into operation. A tiny asteroid, 2004FH, passed within 43,000 km (27,000 miles) of the Earth on 18 March, and the images taken of the fast-moving asteroid enabled astronomers to work out a reliable orbit. In fact 2004FH is not due to hit in the foreseeable future – but others may.

There is another aspect, too. Some objects need to be monitored regularly over long periods, and the conventional great telescopes simply cannot be used in this way. The robotic telescopes can, particularly because they can image many objects, in many different parts of the sky, over the course of a single night. The telescopes will also be used in education; 5 per cent of the time on the La Palma telescope has been allotted to the National School Observatory programme. As was said by Dr Andy Newman, one of the NSO astronomers, 'School children can now work on their own projects alongside professional astronomers. This is the first time that regular access has been granted to schools on a world-class research telescope.' I wonder how many of those schoolchildren will become eminent scientists later on?

Remember, too, that the Liverpool telescopes will have significant advantages over other large telescopes. They operate from unmanned, low-maintenance robotic observatories with clam-shell roofs designed to help with local seeing conditions. All operating costs are cut to a minimum: there is no need to have running water on the site, there are no observatory dormitories and there are no toilets. The telescopes can operate in high winds which would close down any conventional observatory, and objects can be studied with any or all of the five instruments attached to the instrument turret at the Cassegrain focus. All in all the Liverpool telescopes are set to revolutionize time-dependent observational astronomy, and before long there will be six similar telescopes making up the ROBONET worldwide distribution.

It is indeed an important programme, and special congratulations should go to Mike Bode, who planned it as long ago as 1989 and without whom it would never have been started.

33 • PLANETARY NEBULAE

Look upward on a dark summer night in the northern hemisphere of the Earth, and close to the zenith or overhead point in the evening you will see the brilliant blue star Vega, leader of the little constellation of Lyra, the Harp or Lyre. Close to it are two much less conspicuous stars, both of which have individual names – Sheliak and Sulaphat – but are known to astronomers as Beta and Gamma Lyrae. Between these two stars lies a most remarkable object. It is much too faint to be seen with the naked eye; binoculars may show it as a tiny smudge, but a telescope – even a small one – will show you that it looks like a tiny ring. This is what astronomers call a planetary nebula. Many are known, but this one, M57 – the 57th object

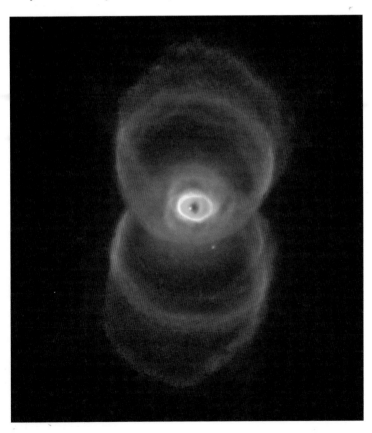

▲ **The Hourglass Nebula,** in the constellation of Musca, imaged by the Hubble Space Telescope. Loops of ejected material surround the central star.

in a famous catalogue drawn up in 1781 by the French astronomer Charles Messier – is one of the brightest, and is certainly the best known.

The name was given to them by the great astronomer Sir William Herschel, who discovered the planet Uranus also in 1781, the year of Messier's catalogue. In fact the name is not a good one, because a planetary nebula is not truly a nebula, and has absolutely nothing to do with a planet. It is in fact a dying star, one that has used up its main store of energy and is close to the end of its luminous career. The planetary nebula stage will not last for long on the cosmic time scale, and little will be left except for a tiny bankrupt star.

Stars in general are long-lived. Our Earth was born around 4,600 million years ago, and the Sun is older than that – 5000 million years is a good estimate, but even so it is no more than middle-aged. It shines not by 'burning' in the conventional sense, but by nuclear reactions going on deep inside it. The main 'fuel' is hydrogen, the most plentiful element in the universe (atoms of hydrogen easily outnumber the atoms of all the other elements combined). Near the Sun's core, where the temperature is of the order of 15 million degrees and the pressure is colossal, hydrogen atoms, or rather nuclei of hydrogen atoms, are combining to make up nuclei of the second-lightest element, helium. It takes four hydrogen nuclei to make one nucleus of helium, and when this happens a little energy is set free; it is this energy that makes the Sun shine. It also means that a little mass (or 'weight', if you like) is lost; the Sun loses 4,000,000 tons every second, so that it weighs much less now than it did when you started to read this page. Fortunately there is plenty of hydrogen left, and the Sun will not change much for at least a thousand million years from now.

No star will last for ever. All are born inside gas-and-dust clouds called nebulae, such as the great nebula M42 in Orion, just below the three stars of the Hunter's Belt (in the northern hemisphere, that is to say; from the southern hemisphere the sword will be above the belt). M42 is 1500 light-years away, so that we see it as it used to be 1500 years ago; it is a veritable stellar nursery, and stars are forming inside it. The nebula is lit up by a multiple star called Theta Orionis, known as the Trapezium because of the arrangement of its four main components. A modest telescope will show the Trapezium easily.

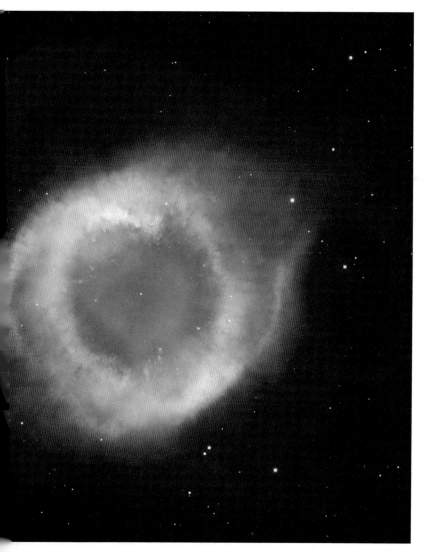

Once a star has been formed, it shines for a very long time, but when the supply of available hydrogen runs low the star has to change. Everything depends upon the initial mass. A star like the Sun will start to use different energy sources, and the inside will shrink and heat up, while the outer layers will expand and cool. The Sun will become a red giant star, and though its surface will have cooled down, its increased size will make it more luminous than it is today. Aldebaran in Taurus, the Bull – not far from Orion in the sky – is a good example of a star in this stage of its evolution. (Betelgeux in Orion is different, because it was of great initial mass, and is now a red 'supergiant', 15,000 times as powerful as the Sun.)

What happens next? The star throws off its outer layers altogether – and turns into a planetary nebula, such as the Ring in Lyra.

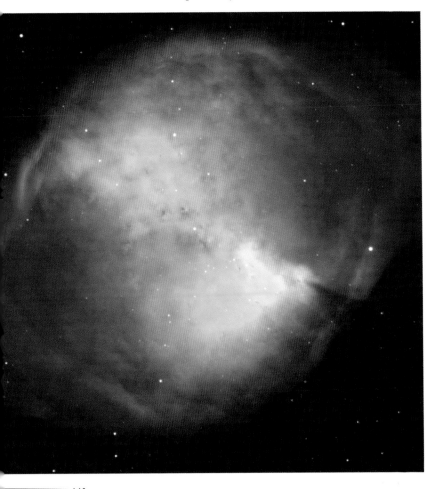

Telescopically it looks beautiful, with its faint but very hot central star, the remnant of the old star. The shapes are varied: the Ring is fairly regular, but other planetary nebulae are not. We have the Helix, the Hourglass, the Dumbbell and many others. The nature of planetaries was established in 1864 by the English amateur William Huggins, who examined the light of an object now known as the Cat's Eye Nebula and found that it was made up of glowing gas. The Cat's Eye was not the first planetary nebula to be found; the first was the Dumbbell Nebula (M27) in the constellation of Vulpecula, the Fox, and the second was the Ring in Lyra, seen by the Frenchman Antoine Darquier in 1779 and shortly afterwards by Messier himself.

The strange forms of some of the planetaries may be explained sometimes by successive outbursts from the dying star; in other cases the star may be one component of a double or binary system. But all planetary nebulae are expanding; the shells of puffed-out material are becoming larger and fainter. Eventually they will fade away altogether, and the material will be dissipated in space. It must be remembered that the gas in a planetary nebula ring is almost incredibly rarefied, and is millions of times less dense than the air we breathe. In fact, it corresponds to what we could call a near-perfect vacuum. When Huggins first analysed the light from the Cat's Eye, he thought that he had found a new element, which he christened nebulium; later, it was found that nebulium is merely oxygen in an unfamiliar state.

◄ **The Dumbbell Nebula,** *M27 in Vulpecula, photographed by the Very Large Telescope. It is one of the most spectacular of the planetary nebulae.*

▼ **The Butterfly Nebula.** *Two lobes of gas expelled by the dying star have created a beautiful winged appearance.*

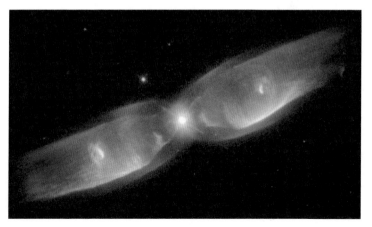

There is little doubt that our Sun will become a planetary nebula several thousands of millions of years hence, but by that time the Earth will no longer be habitable, even if it still exists; it may not survive the Sun's red giant stage, and the two inner planets, Mercury and Venus, will certainly be destroyed. As time passes, the ring will vanish, and all that will be left of the Sun will be a very small body, still shining feebly. This is not the end of the story, and the final fate of the Sun will be its decline into a cold, dead globe – a black dwarf. This will take a very long time indeed, and though the universe as we know it is 13,700 million years old, there may not yet have been enough time for any black dwarfs to have been formed. If they do exist, they will be very hard to detect, because they will send out no energy at all.

The idea of a dead Sun, still circled by the ghosts of its surviving planets, may sound depressing – but at least it will not happen yet awhile, and at least the end will be preceded by a period when any astronomers far away will be able to see our Sun in the guise of a beautiful planetary nebula.

34 • TELESCOPES OF THE ATACAMA DESERT

Where are the best 'seeing' conditions on the surface of the Earth? Different astronomers will have different views. Mauna Kea in Hawaii, Los Muchachos in the Canary Islands and the plateau of the South Pole are favourite candidates, but perhaps the overall vote would go to the Atacama Desert in northern Chile. Here we find several great observatories: La Silla, Cerro Tololo and Las Campanas. *The Sky at Night* has visited all these, and I am indeed glad to have been there. I have also been to Cerro Paranal, where the VLT or Very Large Telescope is now functional. Sadly, I am not now fit enough to go again, so for our programme of November 2004 Chris Lintott went in my place and proved to be a most efficient substitute – many people felt, an improvement!

The altitude is around 3000 metres (10,000 feet), which is not high enough to worry many people, though one has to be careful. The Atacama is rocky, not sandy. At the altitude of the VLT Observatory there is no natural life – no animals, no birds, no insects, no vegetation – only astronomers. Rain falls about once in a hundred years, so that the air is bone-dry, which pleases the radio astronomers, whose

▲ **The Atacama desert;** more specifically, the high-altitude site for the Atacama Large Millimeter Array at Llano de Chajnantor in the Chilean Andes.

◄ *The Melipal* **component,** *part of the Very Large Telescope. It has an 8-metre mirror.*

main enemy is the water vapour in the atmosphere. Of course, there are no towns or villages anywhere near, and the night sky really is inky black.

To be accurate, the VLT is not one telescope but four, each with an 8-metre mirror. When they work together, combining their light, they make up the most powerful telescope ever constructed. The mirrors are probably the most perfect ever made, and two new techniques have been brought into use: active optics and adaptive optics.

A mirror has to move around so as to aim at its target, and this causes distortions, so that something has to be done to keep it in perfect shape. With active optics, computer-controlled pads behind the mirror cope with this, and adjustments can be made with amazing rapidity. But we must also take account of random changes in the atmosphere close to the mirror, and this is where adaptive optics come into the picture. Nothing of the kind would be even remotely possible without modern computers, and it is fair to say that the demands of adaptive optics stretch even the latest computers almost to the limit, but the improvement in image quality is remarkable. Adjustments have to be made hundreds of times in every second.

Another technique is that of interferometry, when the telescopes work together. Interferometry was pioneered by radio astronomers a long time ago but is much more difficult with visible light, because the wavelengths involved are so much shorter. In fact, the VLT was the first telescope to be built specially for interferometric work.

All four main telescopes of the VLT are now in operation, and are soon to be joined by four smaller telescopes which will be mounted on rails, and therefore movable. Being able to alter the configuration means that the VLT can be used for many different research programmes, and this is of vital importance. For example, one astronomer – Alan Fitzsimmons – is studying asteroids, those tiny worlds which are too often dismissed as mere cosmic debris. The rotation periods of NEAs (Near-Earth Asteroids) can be changed by the sunlight falling upon them, and measuring these changes tells us more about what the asteroids are really like; this is important, because there is always the danger of a major collision (as is believed to have happened 65,000,000 years ago, with disastrous results for the dinosaurs). Sooner or later another collision will occur, and we would like to know how we might be able to deal with it. One of these NEAs is 2000 PY5, which is no more than 46 metres (150 feet) across; when best placed it can rise to the 16th magnitude, well within the range of amateur telescopes.

▲ *The domes at La Silla,* *in the southern Atacama Desert. Inaugurated in* *1969, it was the European Southern Observatory's first observing site.*

Most of the ongoing VLT programmes relate to very distant objects, far beyond the Solar System and usually beyond our Galaxy. Here, the immense light-gathering power is paramount. We can examine galaxies 13,000 million light-years away, so that we see them as they used to be when the universe was young. New measurements are being made of the distances of remote galaxies, and of the rate of expansion of the universe. Our main clues come from what are termed Type Ia supernovae; a supernova of this type involves the explosion of the white dwarf component of a binary system. The dwarf pulls material away from its companion, and eventually the situation becomes unstable so that nuclear reactions 'run wild' and blow the dwarf to pieces. All Type Ia supernovae have the same peak

luminosity, and so their distances can be found. Checking these distances against the velocity of recession (found from the Doppler effect in the spectrum) means that we can look back and find how the expansion rate has speeded up, which seems to have happened about 600 million years after the Big Bang creation of the universe. The universe is expanding at an accelerating rate, and will presumably go on doing so until the end of time.

Another investigation involves CDM, or Cold Dark Matter. We now know that visible matter accounts for only a very small fraction of matter in the universe; the rest is invisible, and as yet we have no real idea about its nature. By studying the movements of stars in external galaxies, we can find out how much dark matter is there, and

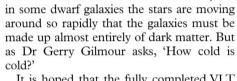

in some dwarf galaxies the stars are moving around so rapidly that the galaxies must be made up almost entirely of dark matter. But as Dr Gerry Gilmour asks, 'How cold is cold?'

It is hoped that the fully completed VLT may be powerful enough to see planets of other stars. We know that these planets exist but so far we have not been able to see them directly. Even if the VLT cannot achieve this, a new telescope – OWL, the Overwhelmingly Large Telescope – may do the trick. At present OWL is in the early planning stage, but it is scheduled to have a 100-metre mirror made up of thousands of individual segments, mounted against each other to give the correct figure (a technique that has been well tested). OWL will weight 14,000 tons – after all, it will be as big as the Jodrell Bank radio telescope – and the engineering problems will be severe by any standards. A site has not yet been selected. The final choice may well rest between the Atacama Desert and the South Pole.

Another major installation in the desert is ALMA, the Atacama Large Millimetre Array, made up of 64 twelve-metre dishes which can be used in various configurations. The altitude is 4600 metres (15,000 feet),

◀ *The Very Large Telescope,* at Paranal, photographed while under construction in December 1997. The observatory platform for the four unit telescopes can be seen, together with the control building on a smaller platform in the foreground.

making it the highest observatory in the world, so that all those working there have to use oxygen. ALMA will work with the VLT and, ultimately, with OWL. Perhaps this will at least show us planets of other stars – even planets like the Earth capable of supporting our kind of life.

So where will all this research lead us? Time will tell. Making forecasts is always a dangerous business, and all I am prepared to say is that I confidently expect the unexpected!

35 · TOUCHDOWN ON TITAN

On 14 January 2005, the Huygens spacecraft made a safe landing on Saturn's largest satellite, Titan. The probe was well named; Titan had been discovered in 1656 by the Dutch observer Christiaan Huygens. It had been 7 years on its way; it had been carried by the Cassini spacecraft, named after the Italian astronomer who had discovered the main division in Saturn's ring system as well as four of the satellites (Iapetus, Rhea, Tethys and Dione).

On Christmas Day 2004 Huygens had been released from Cassini. After that it was unpowered, and no further adjustments were possible. It glided down towards Titan, entering the top of the dense, nitrogen-rich atmosphere at an altitude of 12,860 km (7990 miles) above the surface, at a speed of 18,000 km/h (11,200 mph). The total descent took $2\frac{1}{2}$ hours, during which time analyses of the atmosphere were undertaken and sounds recorded by the sensitive microphone. Using a sequence of parachutes, the speed was dropped to 4300 km/h (2700 mph), and then to 310 km/h (190 mph). When the altitude dropped to 160 km (100 miles) the instruments were exposed to the atmosphere, and the main parachute was replaced by a smaller one. A 20-watt searchlight was switched on to illuminate to ground below; the intensity of sunlight on Titan's surface was only 1/1000 of that on the Earth at noon. The final touch-down speed was no more than 18 km/h (11 mph).

It was a strange scene. The prevailing colour of both ground and sky was orange. It was an immense relief when the first signals were received; up to that time nobody had been sure whether Huygens would land on ice or rock, or splash down in a chemical ocean. In fact, the landing was on the thin crust, and Huygens sank down several inches into spongy, hydrocarbon material with the consistency of wet sand. The heat from the batteries made the methane on the surface boil, and there were puffs of methane gas. Everything had gone well; bear in mind that the whole manoeuvre had been entirely automatic.

▲ *The surface of Titan, as recorded by the Huygens lander. The rocks could be composed of frozen water and hydrocarbons.*

It had been hoped that Huygens would transmit for a few minutes after landing – possibly half an hour. In fact, data were received for 72 minutes, and were relayed back to Earth by the Cassini spacecraft orbiting Saturn. When Cassini passed below the horizon of the landing site, faint carrier signals were still received by radio telescopes in Holland and Germany, and continued for several minutes more. It was equivalent to picking up the signals from a mobile phone over a range of almost a thousand million miles.

Titan is a world of methane rain, rivers often flowing with liquid methane, evaporation lakes, and water ice 'volcanoes'. There are bright icy areas, with snaking and branching riverbeds and large ice blocks; the hills can rise to about 100 metres (300 feet), but there seem to be no high mountains and no impact craters (craters would soon be eroded away by the winds in the thick atmosphere). As expected, the temperature was −180°C. Undoubtedly there is liquid methane no more than a few centimetres below the surface. Whether there are extensive chemical oceans remains to be seen; after all, Huygens could only survey a very limited area, and Cassini has not, as yet, been able to tell whether or not such oceans exist. There is, however, evidence of a coastline, and there may be at least methane lakes.

Icy pebble-sized objects lay near the grounded Huygens; the site was dry, but quite obviously it had been wet not long before, and rain must be very common – not water rain, of course, but methane rain. According to one of the principal investigators, Dr Martin Tomasko of the University of Arizona, the 'mud' in which Huygens landed is made up of 'a mixture of sand, methane and complex organic molecules which form in the upper atmosphere. This smog falls out of the atmosphere and settles on everything. Then methane rain comes, washes it off the ridges and into rivers, then out into the broad

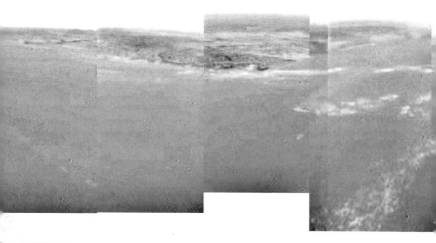

plains, where the rain settles on to the ground and dries up. We are seeing evidence of Earth-like processes, but with very exotic materials.' It also seems that the atmosphere may be similar to the Earth's atmosphere about 4000 million years ago.

We must ask whether there is any chance of life on Titan. On the surface, the answer must be 'no', but there may be oceans up to 100 km (60 miles) deep, made up of water together with 15 per cent ammonia, at a depth of perhaps 310 km (190 miles). There just could be a link here with the methane in the atmosphere. Methane is being

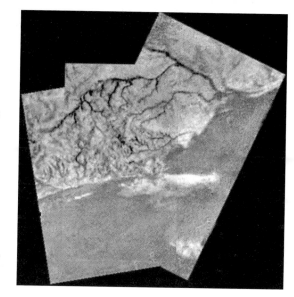

▶ **Riverbeds and lakebeds on Titan,** *imaged by Huygens. Though dry when the lander arrived, the rivers and lakes are thought to have contained a flowing liquid – probably methane – in the recent past.*

▼ **Titan,** *in a composite 360-degree view taken by Huygens from about 8 km (5 miles) above the surface. The white streaks are thought to be ground fog.*

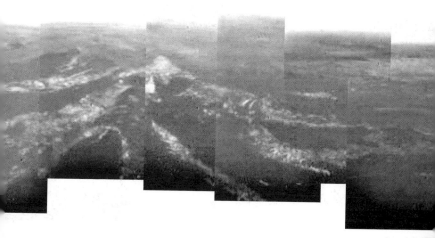

constantly broken up by ultraviolet light from the Sun, so that there must be some source of replenishment from inside the globe. If there is a sub-crustal ocean, methane-producing organisms could have appeared there, and the gas could seep through the surface. Of course, this is only one of several theories, but life has the habit of appearing in the most unexpected places (remember our hydrothermal vents!) and we cannot yet say that Titan is completely sterile.

Huygens is dead, but it has been an outstanding triumph. Apart from the loss of one communications channel, this has been a textbook operation. Cassini is still orbiting Saturn, and has a long programme ahead of sending back data about the rings and the satellites, but its main purpose was to deposit the lander safely. Those who had spent so many years in planning were justifiably elated.

It has been suggested that when the Sun becomes a red giant star, as it must eventually do, Titan may heat up and become suitable for life, while Earth will be scorched to a cinder. The trouble here is that increased temperature means that the atmosphere will escape into space. No, we cannot emigrate to Titan, but we have the satisfaction of knowing that we have at least solved some of the puzzles set by this weird world.

36 • THE OUTER
SOLAR SYSTEM

There was a time – not so long ago – when the Solar System seemed to be a well-defined unit. There was one star (the Sun), nine planets together with their satellites, and much less substantial bodies, such as meteoroids and asteroids, plus the decidedly wraith-like comets. The main planetary system was divided into two well-defined parts, an inner (Mercury to Mars) and an outer (the four giant planets plus Pluto). Pluto was the outermost of the true planets, and beyond it came a vast expanse of space separating us from the nearest night-time star, Proxima Centauri, at over 4 light-years. 'Planet X', moving round the Sun well outside the orbit of Pluto, was regarded as a possibility, but searches for it had been fruitless.

True, the status of Pluto was uncertain. It was found to be smaller than the Moon, and to have a strange orbit; at perihelion it was closer-in than Neptune, as was the case between 1979 and 1999. Not until the very end of the 20th century did Pluto regain its status as 'the outermost planet'.

All the main-belt asteroids, between the orbits of Mars and Jupiter, are small; only Ceres is over 640 km (400 miles) in diameter. In the post-war period Kenneth Edgeworth suggested that there might be a belt of asteroid-sized bodies around or beyond the orbit of Pluto, and a more positive proposal was made a few years later by G.P. Kuiper, but for decades no such bodies turned up.

The first KBO or Kuiper Belt Object was discovered in 1994 by David Mewitt and Jane Luu, observing from the great observatories in Hawaii. The first to be discovered was catalogued as 1994 QB1;

▶ *Comet Ikeya–Zhang, in April 2002, making a return to perihelion for the first time since 1661. It has a longer period than any other comet seen at more than one return.*

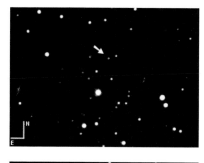

▲ Kuiper Belt Object QB1,
*imaged by John Fletcher, and marked
with an arrow. It has moved against
the stellar background between the
two exposures.*

strangely it has never been given a
proper name, but before long other
small bodies in the far reaches of the
Solar System were found, making
up what we call the Kuiper Belt –
though to be fair it really ought to
be the Edgeworth–Kuiper Belt. By
now hundreds are known, and some
of them are large. Varuna is at least
900 km (550 miles) in diameter;
Quaoar is around 1250 km (800
miles) – and remember that Pluto is
a mere 2300 km (1400 miles)
across. So can there be any real jus-
tification for continuing to class
Pluto as a planet?

Arguments have been going on
about this ever since the large KBOs
were found. The whole matter was
referred in 2000 to the relevant
Commission of the International
Astronomical Union, of which I was
then a member (sadly, I have since
had to retire from it because I am no
longer fit enough to attend meetings
abroad). The decision then was that
Pluto ought to retain its planetary
status, and the proposal to give it an
asteroid number was rejected. Bear in mind, too, that Pluto has a
satellite of its own – Charon, over 1100 km (700 miles) across – and
Pluto has an extensive, though very thin, atmosphere.

The situation has been further complicated by the discovery of a
remarkable body which has been named Sedna, after an Inuit sea
goddess. It stays well outside the known Kuiper Belt; its distance
from the Sun ranges between about 11,000 million km (7000 million miles) and 150,000 million km (93,000 million miles) – and
the revolution period is 12,300 years. It is due at perihelion in the
year 2076, and when it draws back into the far part of its orbit it
will be much too faint to be seen with our telescopes even though
its diameter may be as much as 1900 km (1200 miles), comparable
with Pluto. Its surface seems to be covered with a dark, tarry substance called tholin, and in colour it is very red. At its furthest it lies
far beyond any other known member of the planetary system, and
there have even been suggestions that it may have been captured
from the system of another star.

There may well be other bodies 'out there' which are as large as
Quaoar, Sedna and for that matter, Pluto. The numbers of known

KBOs are increasing rapidly, and some members of the zone can be imaged with small telescopes. John Fletcher has recorded many of them with his 10-inch telescope at Tuffley, Gloucestershire.

And Planet X, a much larger body – does it exist? It may do so, but the evidence is less strong that it seemed to be a few years ago. If it is really there, it will presumably be found one day, but for the moment all we can do is wait and see. Certainly there is a great deal that we still do not know about the outermost part of the Sun's kingdom.

37 • STAR PARTY

'Star parties' are often held in the United States; amateurs (and some professionals) meet at a chosen site and time, so that they can compare notes and also have a very pleasant social event. Such parties are less common in Britain, no doubt because of the vagaries of our climate, but in March 2005 we decided to hold our first *The Sky at Night* event of this kind. We sent out invitations, and at the appointed time we met at my observatory at Selsey in Sussex.

Fortunately Nature was kind. Of the two nights, one was partially clear and the other completely so. We began on Selsey beach (a few hundred metres from my house), because there is a clear sea horizon, and we hoped to catch that elusive planet Mercury, then at its best. Mercury obliged, and so did the extremely thin crescent Moon. Then we made our way back to my home, and set up equipment.

I had three telescopes in action: my 15-inch and 12½-inch reflectors, and my 5-inch Cooke refractor. All were in constant use, but of course many of those taking part brought telescopes and cameras of their own, and it was really good to see how they compared. I admit that my own equipment looks rather dated now, but it has always suited me, as a lunar and planetary observer. The dome of

▲ **Telescopes, large and small,** were set up in my garden at Selsey. The weather was kind, as can be seen in this general view taken by Pete Lawrence.

▶ **Star Party.** The young Moon, with Earthshine, and Mercury, by Pete Lawrence.

◄ **During our Star Party,** the Moon was very young – just a narrow crescent – as can be seen in this photograph by Pete Lawrence.

my 15-inch does look rather like an oil drum!

Of course electronic devices were very much to the fore, and the advantages over conventional photography are indeed startling. Using my telescope with CCD equipment, Damian Peach and Martin Mobberley, for instance, produced pictures of Jupiter and Saturn which were better than any which would have been obtainable at the best professional observer level in 1957, when I presented the very first *Sky at Night* programme. Images of clusters and nebulae have shown equal improvement. All of which goes to show that amateur astronomy is every whit as important now as it was before the setting up of today's great observatories.

Both from a technical and a social aspect, the March star party was a great success. I whole-heartedly recommend functions of this kind. And if I remain with *The Sky at Night* for a year or two yet, I can assure you that our first Star Party will certainly not be our last.

INDEX

Page numbers in *italics* refer
to captions.

More titles from the Philip's Astronomy range

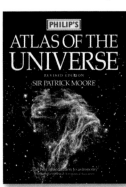

Philip's Atlas of the Universe

Sir Patrick Moore
ISBN 0 540 08791 2   £25.00

- Our Solar System and its place in the Universe

- Includes a complete atlas of the constellations and a Moon map

- How to choose and use binoculars and telescopes

'... *claims to deliver the cosmos and succeeds spectacularly*' Times Literary Supplement

Philip's Planispheres

Rotate the disc to reveal the stars visible from your location at any time on any night of the year. The 51.5°N and 35°S editions come with a season-by-season guide to exploring the skies.

Planisphere 51.5°N
- UK, Northern Europe, Canada
- ISBN 0 540 08817 X • £7.99

Also available 42°N (Southern Europe, USA, Japan), 32°N (Middle East, Northern Africa), 23.5°N (Mexico, Caribbean, India) and 35°S (Australia, New Zealand, South America, Southern Africa).

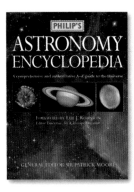

Philip's Astronomy Encyclopedia

General Editor
Sir Patrick Moore
ISBN 0 540 07863 8   £30.00

- The best A–Z encyclopedia for enthusiastic skywatchers

- More than 3000 A–Z articles on all aspects of astronomy and astrophysics

Also available...

The Sky at Night 1992–2001
Sir Patrick Moore • ISBN 0 540 07959
£9.99

Astronomy Dictionary
ISBN 0 540 08689 4 • £9.99

Astrophotography
HJP Arnold • ISBN 0 540 08312 7 • £9

Deep Sky Observer's Guide
Neil Bone • ISBN 0 540 08585 5 • £9.

Mars Observer's Guide
Neil Bone • ISBN 0 540 08387 9 • £8.

Moon Observer's Guide
Peter Grego • ISBN 0 540 08419 0
£9.99

Solar System Observer's Guide
Peter Grego • ISBN 0 540 08827 7
£9.99

Sun Observer's Guide
Pam Spence • ISBN 0 540 08393 3
£9.99

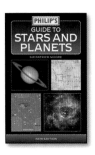

Philip's Guide to Stars & Planets

Sir Patrick Moore
ISBN 0 540 08477 8   £9.99 PB

- Excellent introduction to astronomy for practical observers

- Complete set of constellation maps

- Full colour throughout, with the latest images from the world's best telescopes